# MORE MATH PUZZLES AND GAMES

by
**Michael Holt**

ILLUSTRATIONS BY PAT HICKMAN

WALKER AND COMPANY
New York

Copyright © 1978 by Michael Holt

All rights reserved. No part of this book may be reproduced or transmitted in any form or by any means, electric or mechanical, including photocopying, recording, or by any information storage and retrieval system, without permission in writing from the Publisher.

First published in the United States of America in 1978 by the Walker Publishing Company, Inc.

This paperback edition first published in 1983

ISBN: 0-8027-7229-3

Library of Congress Catalog Card Number: 77-75319

Printed in the United States of America

10 9 8 7 6 5 4 3 2 1

# CONTENTS

Introduction                                            VII

1. Flat and Solid Shapes                                  1
2. Routes, Knots, and Topology                           17
3. Vanishing-Line and Vanishing-Square Puzzles           33
4. Match Puzzles                                         41
5. Coin and Shunting Problems                            49
6. Reasoning and Logical Problems                        56
7. Mathematical Games                                    66

Answers                                                  88

# INTRODUCTION

Here is my second book of mathematical puzzles and games. In it I have put together more brainteasers for your amusement and, perhaps, for your instruction. Most of the puzzles in this book call for practical handiwork rather than for paper and pencil calculations—and there is no harm, of course, in trying to solve them in your head. I should add that none call for practiced skill; all you need is patience and some thought.

For good measure I have included an example of most types of puzzles, from the classical crossing rivers kind to the zany inventions of Lewis Carroll. As with the first book of mathematical puzzles, I am much indebted to two great puzzlists, the American Sam Loyd and his English rival Henry Dudeney.

Whatever the type, however, none call for special knowledge; they simply require powers of deduction, logical detective work, in fact.

The book ends with a goodly assortment of mathematical games. One of the simplest, "Mancalla," dates back to the mists of time and is still played in African villages to this day, as I have myself seen in Kenya. "Sipu" comes from the Sudan and is just as simple. Yet both games have intriguing subtleties you will discover when you play them. There is also a diverse selection of match puzzles, many of which are drawn from Boris A. Kordemsky's delightful ***Moscow Puzzles: Three Hundred Fifty-Nine Mathematical Recreations*** (trans. by Albert Parry, New York: Charles Scribner's Sons, 1972); the most original, however, the one on splitting a triangle's area into three, was given me by a Japanese student while playing with youngsters in a playground in a park in London.

A word on solving hard puzzles. As I said before, don't give up and peek at the answer if you get stuck. That will only spoil the fun. I've usually given generous hints to set you on the right lines. If the hints don't help, put the puzzle aside; later, a new line of attack may occur to you. You can often try to solve an easier puzzle similar to the sticky one. Another way is to guess trial answers just to see if they make sense. With luck you might hit on the right answer. But I agree, lucky hits are not as satisfying as reasoning puzzles out step by step.

If you are really stuck then look up the answer, but only glance at the first few lines. This may give you the clue you need without giving the game away. As you will see, I have written very full answers to the harder problems or those needing several steps to solve, for I used to find it baffling to be greeted with just the answer and no hint as to how to reach it.

However you solve these puzzles and whichever game takes your fancy, I hope you have great fun with them.

—Michael Holt

# 1. Flat and Solid Shapes

All these puzzles are about either flat shapes drawn on paper or solid shapes. They involve very little knowledge of school geometry and can mostly be solved by common sense or by experiment. Some, for example, are about paper folding. The easiest way to solve these is by taking a sheet of paper and folding and cutting it. Others demand a little imagination: You have to visualize, say, a solid cube or how odd-looking solid shapes fit together. One or two look, at first glance, as if they are going to demand heavy geometry. If so, take second thoughts. There may be a perfectly simple solution. Only one of the puzzles is *almost* a trick. Many of the puzzles involve rearranging shapes or cutting them up.

## Real Estate!

K.O. Properties Universal, the sharpest realtors in the West, were putting on the market a triangular plot of land smack on Main Street in the priciest part of the uptown shopping area. K.O.P.U.'s razor-sharp assistant put this ad in the local paper:

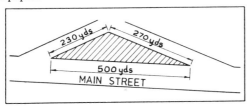

THIS VALUABLE SITE IDEAL FOR
STORES OR OFFICES
Sale on April 1

Why do you think there were no buyers?

## Three-Piece Pie

How can you cut up a triangular cranberry pie this shape into three equal pieces, each the same size and shape? You can do it easily. First cut off the crust with a straight cut and ignore it.

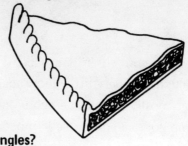

## How Many Rectangles?

How many rectangles can you see?

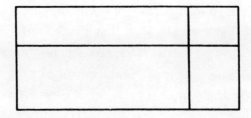

## Squaring Up

How many squares can you find here? Remember, some squares are part of other bigger squares.

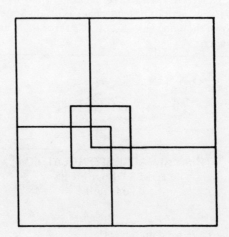

## Triangle Tripling

Copy the blank triangle shown here. Divide it into smaller ones by drawing another shaded triangle in the middle; this makes 4 triangles in all. Then repeat by drawing a triangle in the middle of each of the blank triangles, making 13 triangles altogether. Repeat the process. Now how many shaded and blank triangles will you get? And can you see a pattern to the numbers of triangles? If you can, you will be able to say how many triangles there will be in further divisions without actually drawing in the triangles.

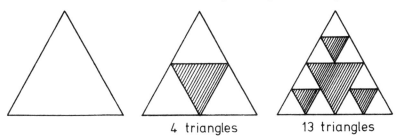

4 triangles       13 triangles

## The Four Shrubs

Can you plant four shrubs at equal distances from each other? How do you do it?

HINT: A square pattern won't do because opposite corners are further apart than corners along one side of the square.

## Triangle Teaser

It's easy to pick out the five triangles in the triangle on the left. But how many triangles can you see in triangle *a* and in triangle *b*?

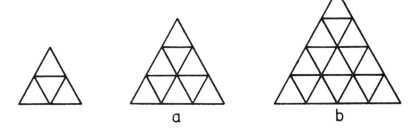

a       b

## Triangle Trickery

Cut a three-four-five triangle out of paper. Or arrange 12 matches as a three-four-five triangle (3 + 4 + 5 = 12).

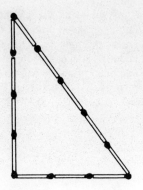

Those of you who know about Pythagoras's theorem will also know it must be right-angled. The Egyptian pyramid builders used ropes with three-four-five knots to make right angles. They were called rope stretchers. The area shut in by the triangle is (3 × 4)/2. If you don't know the formula for the area of a triangle, think of it as half the area of a three-by-four rectangle. The puzzle is this:

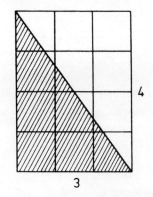

Using the same piece of paper (or the same 12 matches), show 1/3 of 6 = 2.

HINT: This is a *really* difficult puzzle for adults! Think of the triangle divided into thirds this way:

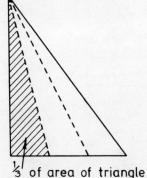

⅓ of area of triangle

If you are using paper, fold it along the dotted lines.

## Fold 'n Cut
Fold a sheet of paper once, then again the opposite way. Cut the corner, as shown. Open the folded sheet out and, as you see, there is one hole, in the middle.

  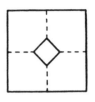

Now guess what happens when you fold three times and cut off the corner. How many holes will there be now?

## Four-Square Dance
How many different ways can you join four squares side to side? Here is one way. Don't count the same way in a different position, like the second one shown here, which is just the same as the first. Only count *different* shapes.

 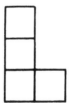

## Net for a Cube
Each shape here is made up of six squares joined side to side. Draw one, cut it out, and it will fold to form a cube. Mathematicians call a plan like this a net. How many different nets for a cube can you draw? Only count *different* ones. For instance, the second net is the same as the first one turned round.

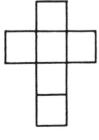 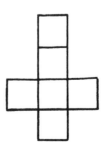

## Stamp Stumper

Phil A. Telist had a sheet of 24 stamps, as shown. He wants to tear out of the sheet just 3 stamps but they must be all joined up. Can you find six *different* ways Phil can do so? The shaded parts show two ways.

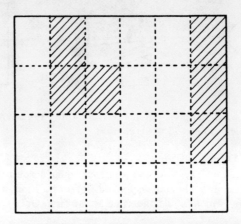

## The Four Oaks

A farmer had a square field with four equally spaced oaks in it standing in a row from the center to the middle of one side, as shown. In his will he left the square field to his four sons "to be divided up into four identical parts, each with its oak." How did the sons divide up the land?

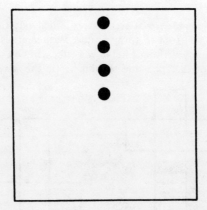

## Box the Dots
Copy this hexagon with its nine dots. Can you draw nine lines of equal length to box off each dot in its own oblong? All oblongs must be the same size, and there must be no gaps between them.

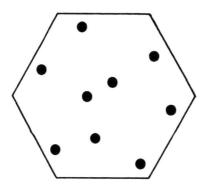

## Cake Cutting
Try to cut the cake shown into the greatest number of pieces with only five straight cuts of the bread knife.

HINT: It's more than the 13 pieces shown here.

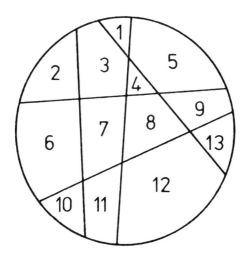

## Four-Town Turnpike

Four towns are placed at the corners of a ten-mile square. A turnpike network is needed to link all four of the towns. What is the shortest network you can plan?

## Obstinate Rectangles

On a sheet of squared paper, mark out a rectangle one square by two squares in size, like this:

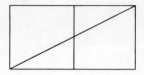

Join a pair of opposite corners with a line, a diagonal. How many squares does it slice through? As you see, two squares. Do the same for a bigger rectangle, two by three squares say. The diagonal cuts four squares.

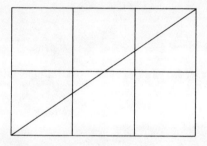

PUZZLE: Can you say how many squares will be cut by the diagonal of a rectangle six by seven squares—without drawing and counting? In short, can you work out a rule? Be careful to work only with rectangles, not squares. It's much harder to find a rule for squares. Stick to rectangles!

HINT: Add the length and the width of each rectangle. Then look at the number of squares cut.

## One Over the Eight

Here is an interesting pattern of numbers you can get by drawing grids with an odd number of squares along each side. Begin with a three-by-three grid, as shown in picture *a*. The central square is shaded, and there are eight squares around it. We have, then, one square in the middle plus the other eight, or 1 + (8 × 1) = 9 squares in all. Now look at grid *b*: It has one central square, shaded, and several step-shaped jigsaw pieces, each made up of three squares. By copying the grid and shading, can you find how many jigsaw pieces make up the complete grid? Then the number of squares in the complete grid should be the number in each "jig" times 8, plus 1: 1 + (8 × 3) = 25. Next, in grid c see if you can copy and finish off the jigsaw pieces; one has been drawn for you. Then complete the number pattern: 1 + 8 jigs = 49. You've got to find what number of squares there are in a jig. Could you write the number pattern for a nine-by-nine grid—without even drawing it?

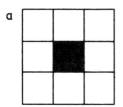

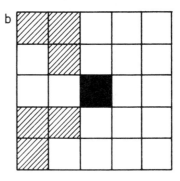

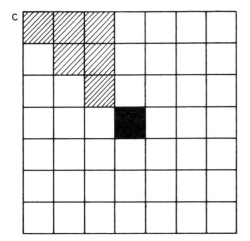

## Greek Cross into Square

Out of some postcards cut several Greek crosses, like these shown here. Each, as you can see, is made up of five squares. What you have to do is cut up a Greek cross and arrange the pieces to form a perfect square. The cuts are indicated on drawings *a*, *b*, and *c*. In the last two puzzles, *d* and *e*, you need two Greek crosses to make up a square. See if you can do it. There is no answer.

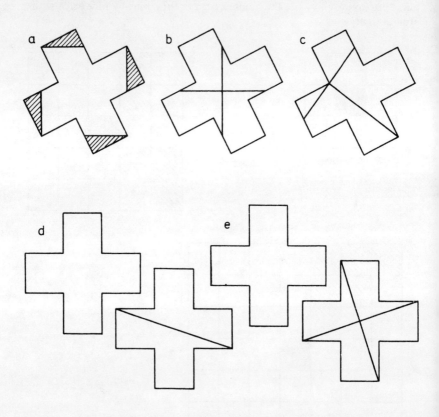

## Inside-out Collar

Take a strip of stiff paper and make it into a square tube. A strip one inch wide and four inches long—with a tab for sticking—will do nicely. Crease the edges and draw or score the diagonals of each face before sticking the ends of the strip together; scissors make a good scoring instrument.

The trick is to turn the tube inside out without tearing it. If you can't do it, turn to the answer section.

## Cocktails for Seven

The picture shows how three cocktail sticks can be connected with cherries to make an equilateral triangle. Can you form seven equilateral triangles with nine cocktail sticks? You can use matchsticks and balls of plasticine instead.

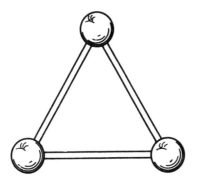

## The Carpenter's Colored Cubes

A carpenter was making a child's game in which pictures are pasted on the six faces of wooden cubes. Suddenly he found he needed twice the surface area that he had on one big cube. How did he double the area without adding another cube?

## Painted Blocks

The outside of this set of blocks is painted. How many square faces are painted?

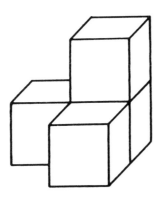

## Instant Insanity

This is a puzzle of putting four identically colored cubes together in a long block so no adjacent squares are the same color. You can make the cubes yourself from the four nets shown in the picture.

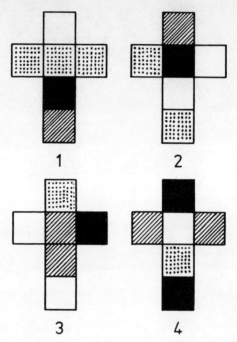

In this puzzle you have four cubes. Each cube's faces are painted with four different colors. Put the four cubes in a long rod so that no colors are repeated along each of the rod's four long sides.

Since there are over 40,000 different arrangements of the cubes in the rod, trying to solve the puzzle in a hit-or-miss fashion is likely to drive you insane!

You can make the cubes yourself by cutting out the four cross-shaped nets shown here. You can, of course, use red, green, blue, and white, for instance, instead of our black, dotted, hatched, and white.

There is a 1-in-3 chance of correctly placing the first cube, which has three like faces. The odds of correctly placing each of the other cubes is 1 in 24: Each cube can be sitting on any of its six faces; and for each of these positions it can be facing the adjacent cube in four different ways—a total of 24 positions. Multiply 3 × 24 × 24 × 24, and the answer is 41,472 —the total number of ways of arranging the cubes. See answer section for solution.

## The Steinhaus Cube

This is a well-known puzzle invented by a mathematician, H. Steinhaus (say it *Stine-house*). The problem is to fit the six odd-shaped pieces together to make the big three-by-three-by-three cube shown at top left of the picture. As you can see, there are three pieces of 4 little cubes and three pieces of 5 little cubes, making 27 little cubes in all—just the right number to make the big cube.

To solve the puzzles, the best thing is to make up the pieces by gluing little wooden cubes together.

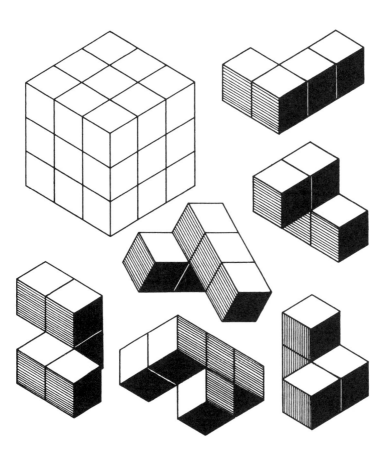

## How Large Is the Cube?
Plato, the Greek philosopher, thought the cube was one of the most perfect shapes. So it's quite possible he wondered about this problem: What size cube has a surface area equal (in number) to its volume? You had better work in inches; of course, Plato didn't!

## Plato's Cubes
A problem that Plato really did dream up is this one: The sketch shows a huge block of marble in the shape of a cube. The block was made out of a certain number of smaller cubes and stood in the middle of a square plaza paved with these smaller marble cubes. There were just as many cubes in the plaza as in the huge block, and they are all precisely the same size. Tell how many cubes are in the huge block and in the square plaza it stands on.

HINT: One way to solve this is by trial and error. Suppose the huge block is 3 cubes high; it then has 3 × 3 × 3, or 27, cubes in it. But the plaza has to be surfaced with exactly this number of cubes. The nearest size plaza is 5 by 5 cubes, which has 25 cubes in it; this is too few. A plaza of 6 by 6 cubes has far too many cubes in it. Try, in turn, a huge block 2, then 4, then 5 blocks high.

## The Half-full Barrel
Two farmers were staring into a large barrel partly filled with ale. One of them said: "It's over half full!" But the other declared: "It's more than half empty." How could they tell without using a ruler, string, bottles, or other measuring devices if it was more or less than exactly half full?

## Cake-Tin Puzzle

The round cake fits snugly into the square tin shown here. The cake's radius is 5 inches. So how large must the tin be?

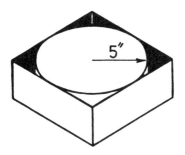

## Animal Cubes

Look at the picture of the dinosaur and the gorilla made out of little cubes. How many cubes make up each animal? That was easy enough, wasn't it? But can you say what the volume of each animal is? The volume of one little cube is a cubic centimeter.

That wasn't too hard, either, was it? All right then, can you say what the surface area of each animal is? The surface area of the face of one little cube is 1 square centimeter.

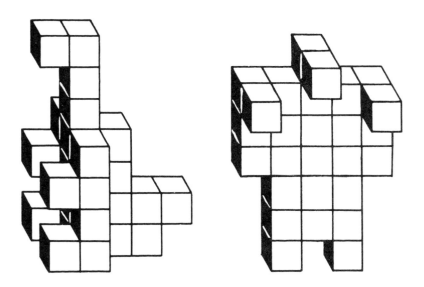

## Spider and Fly
A spider is sitting on one corner of a large box, and a fly sits on the opposite corner. The spider has to be quick if he is to catch the fly. What is his shortest way? There are at least four shortest ways. How many shortest lines can you find?

## The Sly Slant Line
The artist has drawn a rectangle inside a circle. I can tell you that the circle's diameter is 10 inches long. Can you tell me how long the slant line, marked with a question mark, is?

HINT: Don't get tangled up with Pythagoras's theorem. If you don't know it, all the better!

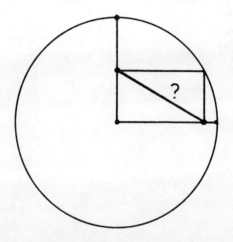

# 2. Routes, Knots, and Topology

In fact all these puzzles are about the math of topology, the geometry of stretchy surfaces. For a fuller description of what topology is about, see the puzzle "The Bridges of Königsberg" on page 25. The puzzles include problems about routes, mazes, knots, and the celebrated Möbius band.

### In-to-out Fly Paths
A fly settles inside each of the shapes shown and tries to cross each side once only, always ending up outside the shape. On which shapes can the fly trace an in-to-out path? The picture shows he can on the triangle. Is there, perhaps, a rule?

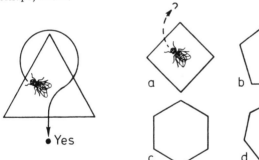

### In-to-in Fly Paths
This time the fly begins *and* ends inside each shape. Can he cross each side once only? The picture shows he cannot do so on the triangle: He cannot cross the third side *and* end up inside. Is there a rule here?

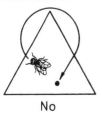

## ABC Maze

Begin at the arrow and let your finger take a walk through this maze. Can you pass along each path once only and come out at *A?* at *B?* and at *C?*

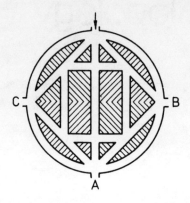

## Eternal Triangle?

Can you draw this sign in one unbroken line without crossing any lines or taking your pencil off the paper? The sign is often seen on Greek monuments. Now go over the same sign in one unbroken line but making the fewest number of turns. Can you draw it in fewer than ten turns?

## The Four Posts
Draw three straight lines to go through the four posts shown here without retracing or lifting your pencil off the paper. And you must return finally to your starting point.

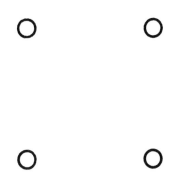

## The Nine Trees
Find four straight lines that touch all nine trees. In this puzzle you don't have to return to your starting point; indeed you cannot! Do the "Four Posts" puzzle and you should be able to do this one.

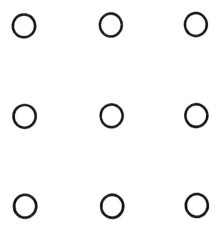

## Salesman's Round Trip

A traveling salesman starts from his home at Anville (*A*). He has to visit all three towns shown on the sketch map—Beeburg (*B*), Ceton (*C*), and Dee City (*D*). But he wants to save as much gas as he can. What is his shortest route? The map shows the distances between each town. So *A* is eight miles from *C*, and *B* is six miles from *D*.

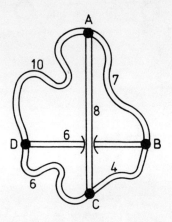

## Swiss Race

The sketch map here shows the roads on a race through the Swiss Alps from Anlaken (*A*) to Edelweiss (*E*) through the checkpoints *B*, *C*, and *D*. An avalanche blocks the roads at three points, as you can see. You've got to clear just one roadblock to make the shortest way to get through from Anlaken to Edelweiss. Which one is it? And how long is the route then?

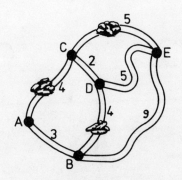

## Get Through the Mozmaze

The maze shown here is called a mozmaze because it is full of awful, biting dogs, called mozzles. Top Cat is at the top left-hand corner, and he has to get through the mozmaze to the lower right corner, where it says END. But on his way he has to pass the biting mozzles chained at the various corners of the mozmaze. The triangles mark the position of the dogs that give three bites as Top Cat passes each of them; the squares of the dogs that give two bites; and the circles of the dogs that give only one bite.

What is Top Cat's best way through the mozmaze so that he gets bitten the fewest times? What's the fewest number of bites he can get by with? Can you do better than 40 bites?

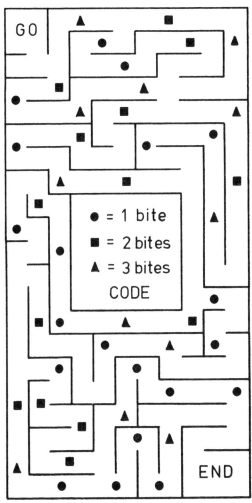

### Space-Station Map

Here is a map of the newly built space stations and the shuttle service linking them in A.D. 2000. Start at the station marked *T*, in the south, and see if you can spell out a complete English sentence by making a round-trip tour of all the stations. Visit each station only once, and return to the starting point.

This puzzle is based on a celebrated one by America's greatest puzzlist, Sam Loyd. When it first appeared in a magazine, more than fifty thousand readers reported, "There is no possible way." Yet it is a really simple puzzle.

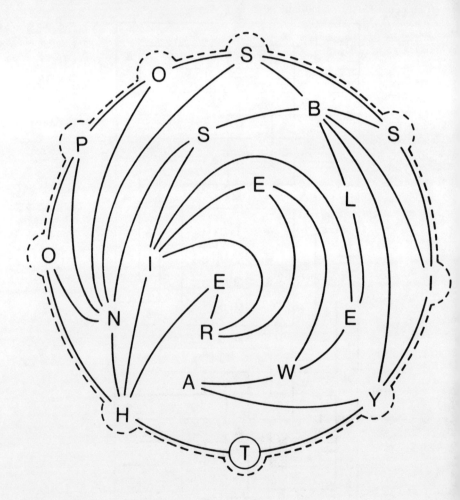

## Round-Trip Flight
Trans-Am Airways offers flight links between these five cities: Albany, Baltimore, Chicago, Detroit, and El Paso. There are eight flights, as follows: Baltimore to Chicago, Detroit to Chicago, Albany to Baltimore, Chicago to El Paso, Chicago to Detroit, Baltimore to Albany, Albany to El Paso, and Chicago to Albany. What is the shortest way to make a trip from Albany to Detroit and back again?

HINT: Draw a sketch map of the flights, beginning: $A \rightleftarrows B \longrightarrow C$. This will show you how to avoid making too many flights *or* getting stuck in a "trap!"

## Faces, Corners, and Edges
Here is a surprising rule about shapes you should be able to puzzle out for yourself. Find a box—a matchbox, a book, or a candy box, say. Now run your finger along the *edges* and count them (12) and add 2 to the number you found (making 14). Now count the number of faces (6) and add to that number the number of corners (8), making 14 in all. It seems that there is a rule here. Count faces and corners and edges of the shapes shown in our picture; the dotted lines indicate hidden edges that you cannot see from the head-on view. Can you find the rule? The great Swiss mathematician Leonhard Euler (say it *oiler*) was the first to spot it. The names of the shapes are *tetrahedron* (4 faces), *octahedron* (8 faces), *dodecahedron* (12 faces), and *icosahedron* (20 faces).

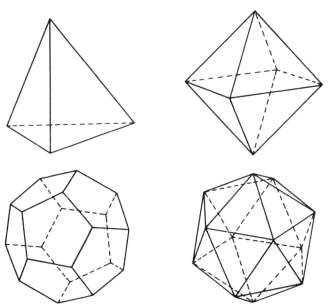

## Five City Freeways

A planner wants to link up five cities by freeways. Each city must be linked to every other one. What's the least number of roads he must have? Roads can cross by means of overpasses, of course.

The planner then decides that overpasses are very costly. What is the fewest number of overpasses he needs?

## The Bickering Neighbors

There were three neighbors who shared the fenced park shown in the picture. Very soon they fell to bickering with one another. The owner of the center house complained that his neighbor's dog dug up his garden and promptly built a fenced pathway to the opening at the bottom of the picture. Then the neighbor on the right built a path from his house to the opening on the left, and the man on the left built a path to the opening on the right. None of the paths crossed.

Can you draw the paths?

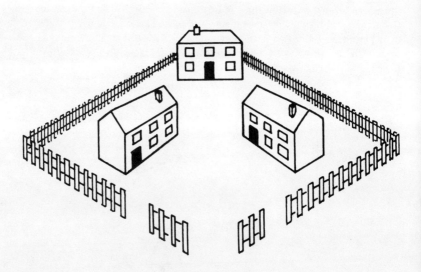

## The Bridges of Königsberg
This is one of the most famous problems in all math. It saw the start of a whole new branch of math called *topology*, the geometry of stretchy surfaces. The problem arose in the 1700s in the north German town of Königsberg, built on the River Pregel, which, as the picture shows, splits the town into four parts.

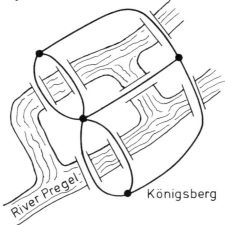

In summer the townsfolk liked to take an evening stroll across the seven bridges. To their surprise they discovered a strange thing. They found they could not cross all the bridges once and once only in a single stroll without retracing their steps. Copy the map of Königsberg if this is not your book, and see if you agree with the Königsbergers.

The problem reached the ears of the great Swiss mathematician Leonhard Euler. He drew a basic network, as mathematicians would say, of the routes linking the four parts of the town. This cut out all the unnecessary details. Now follow the strolls on the network. Do you think the Königsbergers could manage such a stroll or not?

## Euler's Bridges

Euler actually solved the last problem in a slightly different way from the one we gave, which is the way most books give. What he did was to simplify the problem. He started off with the very simple problems we give below. He then went on from their solutions to arrive at the solution we gave to "The Bridges of Königsberg." The little problems go like this:

A straight river has a north bank and a south bank with three bridges crossing it. Starting on the north bank and crossing each bridge once only in one stroll without retracing your steps, you touch the north bank twice (see picture *a*). For five bridges (picture *b*) you touch north three times. Can you find a rule for any odd number of bridges?

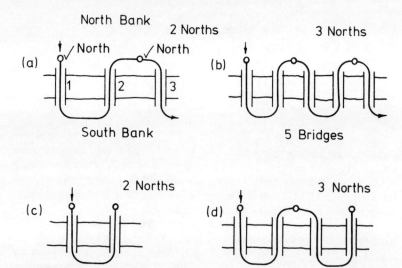

Now look at picture *c*. You touch the north bank twice for two bridges; and as shown by picture *d*, you touch north three times for four bridges. Can you find a rule for any even number of bridges?

## Möbius Band

One of the most famous oddities in topology is the one-edged, single-surfaced band invented by August Möbius. He was a nineteenth-century German professor of math. Take a collar and before joining it give it one half-twist. Now cut it all the way along its middle. How many parts do you think it will fall into? You can try this on your friends as a party trick. Then try cutting it one third in from an edge, all the way round. How many parts do you think it will fall into now?

## Double Möbius Band

Take two strips of paper and place them together, as shown. Give them both a half-twist and then join their ends, as shown in the picture. We now have what seems to be pair of nested Möbius bands. You can show there are two bands by putting your finger between the bands and running it all the way around them till you come back to where you started from. So a bug crawling between the bands could circle them for ever and ever. It would always walk along one strip with the other strip sliding along its back. Nowhere would he find the "floor" meeting the "ceiling." In fact, both floor and ceiling are one and the same surface. What seems to be two bands is actually . . . . Find out and then turn to the answer section to see if you were right. As an added twist, having unnested the band(s), see if you can put it (them) back together again.

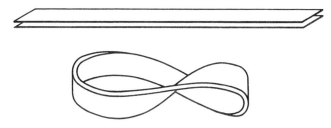

## Viennese Knot

In the 1880s in Vienna a wildly popular magician's trick was to put a knot in a paper strip simply by cutting it with scissors. This is how it was done:

Take a strip of paper, about an inch wide and a couple of feet long. Just before joining the ends, give one end a twist of *one and a half* turns. (If you have read about the Möbius band, you'll know this is like making one with an extra twist in it.) Then tape the ends together to form a band. That done, cut along the middle of the closed band until you come back to where you started. At the last snip you will be left with one long band, which you will find has a knot in it. Pull it and you should see a knot in the shape of a perfect pentagon.

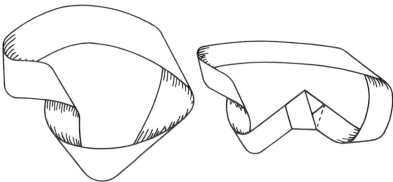

### Release the Prisoners

Here is another problem in topology. Connect your wrists with a longish piece of rope. Make sure the loops around your wrists are not too tight. Have a friend do the same, but before completing the tying up, loop his rope around yours, as shown in the picture.

Can you separate yourself from your friend without untying the knots or cutting the rope? It can be done!

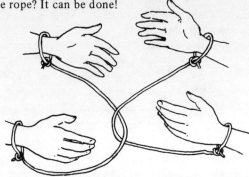

### Three-Ring Rope Trick

This is a famous problem from topology that with a little trial and error I am sure you can solve for yourself. First make three loops of rope or string and link them in a chain like a Christmas decoration. Cut the middle loop and all three pieces of rope will come unlinked. Cut either end loop and the other two stay linked. The puzzle is this: Can you link three loops of rope so that all three will come unlinked if any one is cut? It can be done.

### Wedding Knots

Russian girls use straws to foretell whether they will be married during the year. A girl will take six straws and fold each of them in half, keeping the folds hidden in her fist. Then she asks another girl to tie the 12 straw ends together in pairs; if a complete circle of straws is formed, she will be married within the year.

You can make a closed loop with four straws in two ways, as shown. String will do instead of straws. Can you join the loose ends of six straws to make a single closed loop in three different ways?

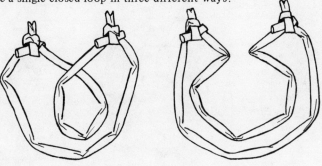

## Amaze Your Friends

Ask a friend to draw a maze with a pencil on a large sheet of paper. He can make it as twisty as he likes, but none of the lines may cross and the ends must join to make a closed loop. Now newspapers are placed around the edges as shown here so that only the middle part of the maze shows. The

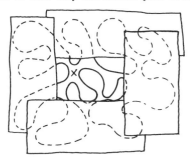

friend now places his finger anywhere in this still exposed area. Is his finger inside or outside the maze? The maze is so complicated it must be impossible to say which points are inside the closed loop and which points are outside. All the same you state correctly whether his finger is inside or outside the maze.

Another way to present the trick is with string or rope. Take a good length and tie the ends to form a long loop. Then ask the friend to make a closed-loop maze with it. Put newspapers down to hide the outside of the maze. The friend puts his finger on some spot in the maze. Take one newspaper away and pull an outside part of the string across the floor. Will the string catch on the friend's finger or not? Again you predict correctly each time the trick is performed. How is it done?

The secret is this: Take two points in the maze and join them with an imaginary line. If the points are both inside the loop, then the line will cross an *even* number of strings. If both points are outside, the same rule holds. But if one point is inside and the other outside, then the line connecting them will cross an *odd* number of points. The easiest way to remember the rule is to think of the simplest maze possible, a circle. If both points are inside the circle (or both outside it), then the line connecting them will cross either no strings or two strings; both 0 and 2 are even numbers. If one point is inside the other outside, then the line will cross the circle once; 1 is an odd number.

To do the stunt, as the newspapers are being placed, let your eye move through the maze from the outside until you reach a spot near the center that is easy to remember. You know that spot is outside the maze. When your friend places his finger, you have only to draw mentally a line from your "outside" spot to his finger and note whether you cross an even number of strings (then his finger is outside) or an odd (his finger is inside). A little practice will show that the trick is easier to do than to describe.

29

## Tied in Knots?

Pull the ends of each rope shown here and find out which will tie itself in a knot. Knot *h* is very interesting; it is often used by magicians. It is known as the Chefalo knot. It is made from the reef knot shown in *g*.

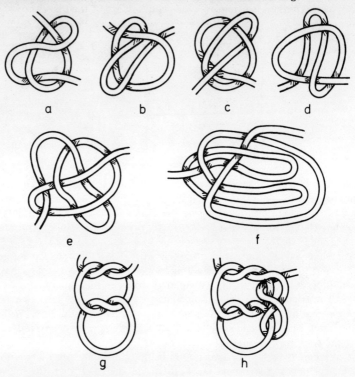

## The Bridges of Paris

In 1618 the plan of Paris and its bridges over the River Seine looked like the sketch map here. The famous Nôtre Dame Cathedral is shown by the † on the island. Could the Parisians then take a stroll over the bridges and cross each one only once without retracing their steps? Draw a network as was done for "The Bridges of Königsberg."

## Tour of the Castle

The idea here is that you have to visit each room in the castle only once on a tour of it, starting at the *in* arrow and leaving by the *out* arrow. With the *exit* placed as in the first of the little 4-roomed castles shown here you can do it; in the second you cannot.

Try your hand at (*1*) the 9-roomed castles, and (*2*) the 16-roomed castles.

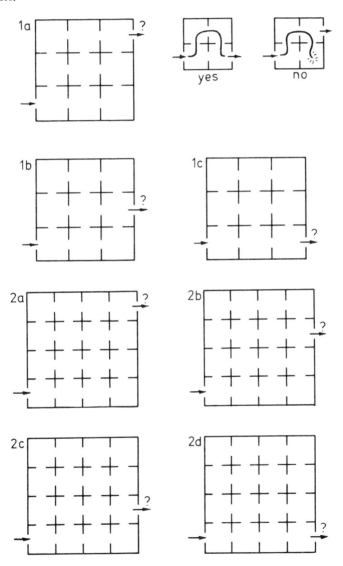

## The Cuban Gunrunners Problem
The Cuban gunrunners plan to transport a trainload of guns and bombs from Havana to Santiago. There are several rail routes they could take, as you can see on the map of the rail system shown. How can they be stopped from getting through? The easiest way is to blow up a few bridges. What is the fewest number of bridges you must blow up? And which ones are they?

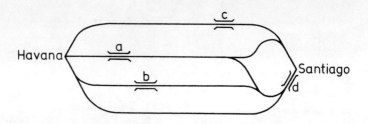

# 3. Vanishing-Line and Vanishing-Square Puzzles

To Martin Gardner, America's leading popularizer of math and purveyor of puzzles, I owe ideas for this next lot of puzzles, all of which, and more, are to be found in his excellent *Mathematics, Magic and Mystery* (New York: Dover, 1956).

These puzzles all depend on a strange quirk of geometry. All but the first involve cutting and rearranging parts of a figure. When that is done, a part of the figure or a line apparently vanishes. Where has it gone? is the question. Before I describe some of these puzzles and explain them, look at the following puzzle, about counting, not about cutting up figures; it gives the clue to the puzzles of the vanished lines.

There are no answers except to the next puzzle.

## Mr. Mad and the Mandarins
Mr. Mad was having three children to tea. Four places were laid, each with three mandarin oranges on a plate. But one of the children didn't turn up. So how should the others divide up the spare plateful of mandarins? Mr. Mad suggested this way, as shown:

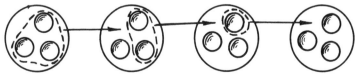

All three mandarins on the first plate went to the second plate, from which two mandarins were put on the next plate, from which one mandarin was placed on the last plate, Mr. Mad's. "There!" exclaimed Mr. Mad. "Fair shares for all. But I bet you can't tell me which plateful has vanished?" None of the children could give an answer. Can you suggest one?

## The Vanishing-Line Trick
Mightily simplified though this puzzle is, it forms the heart of many brilliant puzzles created by Sam Loyd, the great puzzlist. Draw on a card three equal lines, as shown here. Make certain that both the first and the third line touch the diagonal of the card (the broken line), each with one of its ends. Cut the card along the diagonal. Slide the top half to the right until the lines coincide again, as in the second picture. There are now only two lines where before there were three. What has happened to the third line? Which line vanished and where did it go? Slide the top part back and the third line returns.

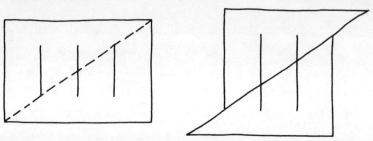

It is like the vanishing group of mandarins in "Mr. Mad and the Mandarins" puzzle. What happens is that the middle line is broken into two parts —one going to lengthen the first line, the other lengthening the third line. With more lines the distribution of the lines is less obvious and the disappearance of the center line becomes even more puzzling.

## The Vanishing-Face Trick
We can doll up "The Vanishing-Line Trick" by drawing pictures instead of lines. Our top picture shows six cartoon faces divided by a broken line into two strips.

Copy and cut out the strips in the top picture and stick each strip on a card. Shift the upper strip to the right, and—as the bottom picture shows—all the hats remain, but one face vanishes. You cannot say which face has vanished. Four of them have been split into two parts and the parts redistributed so that each new face has gained a small bit.

## The Vanishing-Square Trick

Conjurers perform miraculous-seeming tricks where a rectangular or square figure is cut up and rearranged and in the process a whole square is lost to view. The simplest and oldest example explains how it is done. The following explanation is based on Martin Gardner's excellent book *Mathematics, Magic and Mystery*:

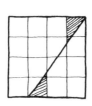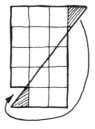

Start with a 4-by-4 square; its area is 16. This square is cut along the slant line. This line is *not* a diagonal, since it passes through only one of the corners. This is the secret of the trick. Now shift the lower part of the board to the left, as shown in the right-hand picture. Snip off the shaded triangle sticking out at the top right corner and fit it into the space at the lower left corner, as shown by the arrow. This produces a 3-by-5 rectangle; its area is 15. Yet we started with a big square of area 16. Where has the missing little square gone? As we said, the secret lies in the way the slant line was drawn. Because that line is not a diagonal, the snipped-off triangle is taller than 1: It is 1⅓ in height. So the rectangle's height is actually 5⅓, not 5. Its actual area is then 3 × 5⅓ = 16. So, you see (or rather you didn't "see"), we haven't lost a square. It just looks like it.

The trick is not very baffling with such a small board. But a larger number of squares will conceal the secret. You can see that this puzzle is like "The Vanishing-Line Trick" when you look at the squares cut by the slant line. As you move up the line, you find that the parts of the cut squares above the line get smaller and smaller while those below get larger and larger—just like the vertical lines in the earlier puzzle.

## Sleight of Square

In "The Vanishing-Square Trick" all the trickery is confined to the squares either side of the slant line. The rest of the square plays no part in the trick at all; it is there merely for disguise. Now instead of cutting the square board into two pieces, suppose we chop it into four. The trick would become even more mysterious. One way to do this is shown in the picture of the 8-by-8 board.

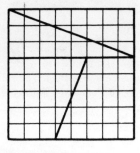
AREA = 8 × 8 = 64

AREA = 5 × 13 = 65

When the four pieces are rearranged, there is a gain of 1 square—from 64 to 65 squares. You'll find there is a long, thin, diamond-shaped gap along the diagonal of the 5-by-13 rectangle. This is hardly noticeable. But it is where the "extra square" has come from. If you were to begin with the 5-by-13 rectangle, drawing an accurate diagonal, then in the 8-by-8 square the upper rectangle would be a shade higher than it should be and the lower rectangle a bit wider. This bad fit is more noticeable than the slight gap along the diagonal. So the first method is better.

Sam Loyd, Jr., discovered how to put the four pieces together to get an area of only 63—that is, to lose a square. This picture shows how it is done.

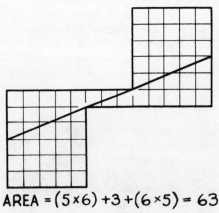
AREA = (5×6) + 3 + (6×5) = 63

## The Secret Fibonacci Lengths

You can make a square come and go at will with other size boards—provided you know the secret lengths of the perpendicular sides (excluding the slant lines) of all pieces—both the inner cutouts and the boards made out of them. In "Sleight of Square" these lengths were 3, 5, 8, and 13. These numbers are part of a famous number series, the Fibonacci (say it *fib-o-NAH-chee*) numbers. It goes 1, 1, 2, 3, 5, 8, 13, 21, 34, and so on. Each number from 2 onward is the sum of the two previous numbers: $3 = 1 + 2$, $5 = 3 + 2$, and so on. Fibonacci, an Italian, was the first great European mathematician; he lived in the 1200s. I doubt if he ever foresaw this curious use of his number series for geometrical trickery!

So we started with an 8-by-8 square with an area of 64 and ended up with a 5-by-13 rectangle with an area of 65. And you notice 8 lines between 5 and 13 in the Fibonacci series.

The trick works with higher numbers in the series; the higher the better because the "extra square" is more easily lost in a longer diagonal. For example, we can choose a 13-by-13 square, with an area of 169, and divide its sides into lengths of 5 and 8, as shown.

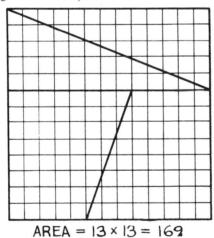

AREA = 13 × 13 = 169

Cutting along the lines, we can rearrange the pieces into an 8-by-21 rectangle, with area 168. A little square has been *lost*, not gained, this time. The Fibonacci numbers used are 5, 8, 13, and 21. There is a loss of a square because the pieces along the diagonal overlap instead of having a gap between them. An odd fact emerges. A board using the lengths 3, 8, 21, and so on—that is, every other Fibonacci number—gives a gain of a square. A board using the lengths 5, 13, 34, and so on results in a loss of a little square.

If you cut up a 2-by-2 board, making a 3-by-1 rectangle, the overlap (resulting in a loss of a quarter of the board) is too obvious. And all the mystery is lost.

## Langman's Rectangle
A rectangle can also be cut up and the pieces fitted together to make a larger rectangle. Dr. Harry Langman of New York City has devised a way of cutting up a rectangle. His method, shown below, makes use of the Fibonacci numbers 2, 3, 5, 8, 13, and 21.

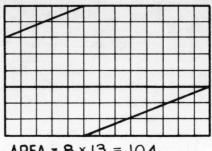

AREA = 8 × 13 = 104

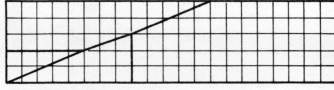

AREA = 5 × 21 = 105

## Curry's Paradox
A paradox is an absurd trick that on the face of it looks flawless. An 11-by-11 square is cut into five pieces, as shown here. The paradox is: When the pieces are put together in another way, a hole appears. Two squares have seemingly been lost. One of the L-shaped pieces must be shifted to produce the effect.

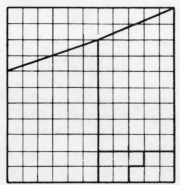
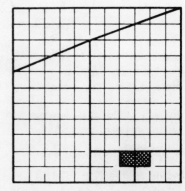

This paradox was invented by a New York City amateur magician Paul Curry in 1953. He also devised a version using a 13-by-13 square where a still larger hole appears and three squares are lost. As you see, Curry's paradox uses Fibonacci numbers.

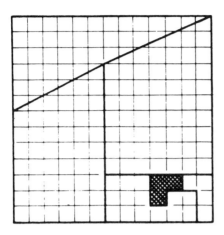

## Gardner's Triangle

It's possible to make part of a triangle disappear. Martin Gardner has applied Curry's paradox to a triangle. His method is shown in the picture of the triangles with two equal sides. By rearranging the six pieces, two squares are lost.

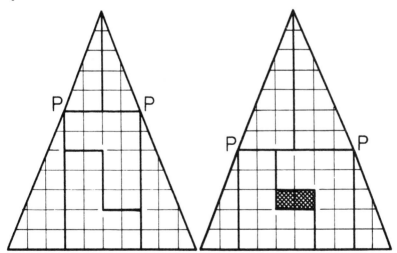

The deception is increased by having the points *P* fall exactly on the crossings of the grid, since the sides will slightly cave in or out.

## Hole in the Square

Another quite different way of losing area is to cut a square into four exactly equal pieces, as shown, by two crosscuts.

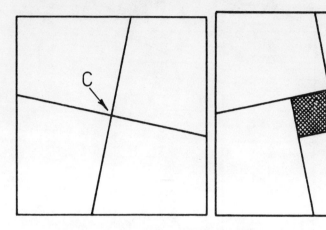

Rearrange the pieces, and a square hole appears in the center. The size of the hole varies with the angle of the cuts. The area of the hole is spread around the sides of the square. This trick suffers from the fact that it is fairly obvious that the sides of the square with the hole are a bit longer than the sides of the first square.

A more mysterious way of cutting a square into four pieces to form a hole is shown in this picture.

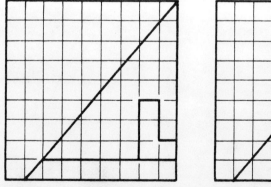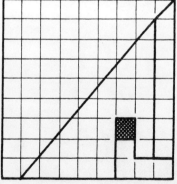

The effect is based on "The Vanishing Square Trick," described earlier. Two of the pieces must be shifted to produce the effect—the long strip of the lower edge and the L-shaped piece. If you remove the largest piece (top left), you are left with another Gardner triangle.

# 4. Match Puzzles

The next batch of puzzles all make use of matches. Toothpicks or orange sticks all the same length are equally good.

## Squares from 24 Matches

Take 24 matches. How many squares the same size can you get with them? With 6 matches to a side you get one square. You can't come out even with squares with 5 and 4 matches to a side. With 3 matches to a side, you get two squares, as shown.

With 2 matches to a side, you get three squares:

Suppose we allow squares of *different* sizes. (*a*) With 3 matches to a side, how many extra, smaller squares can you get now? (*Clue:* Squares can overlap.) (*b*) With 2 matches to a side show how you can get a total of seven squares.

With 1 match on a side you can make six identical squares, as shown.

(*c*) With 1 match on a side how do you make seven identical squares? Eight identical squares? And nine identical squares? There'll be some extra, bigger squares too. With the nine squares there are five extra squares.

41

## PART-MATCH SQUARES
*The next three puzzles need 24 matches. You make half-match squares and other part-match squares by crossing one match over another.*

### Half-Match Squares
Use half a match as the side of a square. Can you get 16 small squares? How many larger squares can you see?

### Third-Match Squares
Can you get 27 small squares, one third of a match on each side? How many larger squares can you see?

### Fifth-Match Squares
Can you get 50 small squares in 2 match-stick size squares? How many larger squares of all sizes can you see?

### Move-or-Remove Puzzles I
Begin with 12 matches, making four small squares as shown.

(a) Remove 2 matches, leaving two squares of different sizes
(b) Remove 4 matches, leaving two equal squares
(c) Move 3 matches to make three squares the same size
(d) Move 4 matches to make three squares the same size
(e) Move 2 matches to make seven squares of various sizes (you'll have to cross one match over another)
(f) Move 4 matches to make 10 squares, not all the same size (you'll have to cross one match over another more than once)

### Move-or-Remove Puzzles II
Begin with 24 matches, making nine small squares as shown.

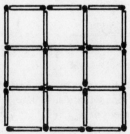

(a) Move 12 matches to make two squares the same size
(b) Remove 4 matches, leaving four small squares and one large square
(c) Remove 6 matches, leaving three squares
(d) Remove 8 matches, leaving four squares, each 1 match to a side (two answers)
(e) Remove 8 matches, leaving two squares (two answers)
(f) Remove 8 matches, leaving three squares
(g) Remove 6 matches, leaving two squares and two L-shaped figures
(h) Remove 4, 6, then 8 matches to make five squares, each 1 match to a side

## Windows
Make six squares—not all the same size—with nine matches. The answer looks like two windows.

## Greek Temple
The temple shown is made out of 11 matches.

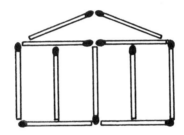

(A) Move 2 matches and get 11 squares
(B) Move 4 matches and get 15 squares

## An Arrow
This arrow is made of 16 matches.

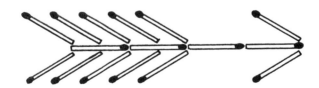

(A) Move 10 matches in this arrow to form eight equal triangles
(B) Move 7 matches to make five equal four-sided figures

## Vanishing Trick
There are 16 squares here with one match on a side. But how many squares in all?

Take away nine matches and make every square—of any size—vanish.

## Take Two
The eight matches here form, as you see, 14 squares.

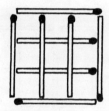

Take two matches and leave only 3 squares.

## Six Triangles
Three matches make an equal-sided, or equilateral, triangle. Use 12 matches to make six equilateral triangles, all the same size. That done, move 4 of the matches to make three equilateral triangles *not* all the same size.

## Squares and Diamonds
Form three squares out of ten matches. Remove one match. Leaving one of the squares, arrange the other five matches around it to make two diamonds.

## Stars and Squares
Put down eight matches to make two squares, eight triangles, and an eight-pointed star. The matches may overlap.

## A Grille
In the grille shown here move 14 matches to make three squares.

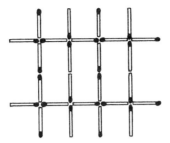

## The Five Corrals
Here is a field, four matches square. In it there is a barn one match square. The farmer wishes to fence off the field into five equal L-shaped corrals. How does he do it? (Use ten more matches for the fencing.)

## Patio and Well
In the middle of this patio, five matches square, is a square well.

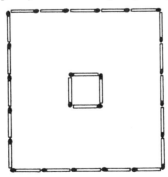

(a) Use 18 more matches to split the patio into six L-shaped tiles all the same size and shape
(b) Use 20 more matches to split the patio into eight equal L-shaped tiles

## Four Equal Plots

Here is a square building site 4 matches on a side. We will call its area 16 square match units (4 × 4 = 16).

Add 11 matches to fence off the site into four plots, each with an area of 4 square match units. But you must do it so that each plot borders on the other three. One of the plots is a square, two are L-shaped, and one is a rectangle.

## Get Across the Pool

Here is a garden pool with a square island in the middle.

Add two "planks" (matches) and step across the water onto the island.

## Spiral into Squares

Move four matches in this spiral in order to form three squares.

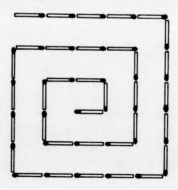

## More Triangle Trickery

Make a three-four-five triangle out of 12 matches. The matches shut in an area of 6 square match units. (This is easy to see because the triangle is exactly half of a three-by-four rectangle, whose area would be 12 square match units.)

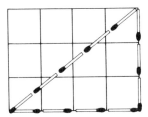

(*a*) Move 3 matches to form a shape with area 4 square match units
(*b*) Move 4 matches to form a shape with area 3 square match units

CLUE: In both *a* and *b* move the matches from the shorter sides of the triangle.

## Triangle Trio

Can you make just three equal-sided triangles out of seven matches?

## Triangle Quartet

With these six matches can you make four equal-sided triangles?

## 3 Times the Area

Look at the rectangle on the left. It has 3 times the area of the rectangle on the right, as the dotted lines show.

Add one match to the smaller rectangle so it has 7 matches. Make it into a box-girder shape made up of three equal-sided triangles. Now add four matches to the rectangle on the left and make it into a shape made up of 19 equal-sided triangles—so it has an area 3 times as great as the box-girder shape.

47

## Cherry in the Glass

Arrange a penny and four matches as shown. This is your cherry in a glass. Take the cherry out of the glass simply by moving two matches. You must not touch the cherry (penny), of course.

# 5. Coin and Shunting Problems

The next section includes puzzles about shuffling coins and the classic puzzles of ferrying people across rivers in boats. Also there is a wide selection of railway shunting problems. The best way to solve these is by actually drawing a plan of railway tracks, making coins or bits of paper stand for the engines and their cars, and moving them about on the tracks. It is a good idea to jot down the moves you make so you don't forget them, particularly if you are successful and solve the problem. There is nothing more vexing than to solve such a puzzle and then not be able to remember your moves!

**Coin Sorting in Pairs**
Arrange three pennies and two dimes in a row, penny-dime-penny-dime-penny. Move the coins in pairs so that the three pennies are together and next to them the two dimes, as shown in the second picture.

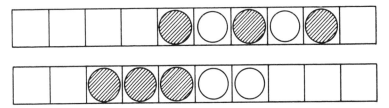

A move is like this: You place the tips of your first and second fingers on any two coins—the coins don't have to be next to each other or of the same denomination—then you slide the pair to another part of the row, but you must keep the same spacing between the coins. It helps to use squares to keep the spacing. You must not make any pair of coins merely change places. When you finish, there must be no spaces between any of the coins. You can move the coins as many spaces as you like left or right. But ten spaces should be enough. Can you do it in three moves?

## Rats in a Tunnel

Two brown rats and two white rats met head on in a tunnel. How did they pass one another and change ends of the tunnel? They could only move by moving forward into a space or by hopping over another rat (of their own or the other color) into a space. Or they could move back. What is the fewest number of moves needed to change the rats over? Here are the kind of moves you can make:

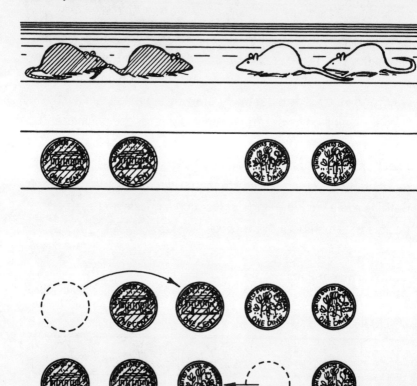

To work it out, use two pennies (for the brown rats) and two dimes (for the white rats). Put them in a line with a gap between, as shown in our sketch.

## Three-Coin Trick

Begin with three coins showing a head placed between two tails. Each move in this puzzle consists of turning over two coins *next* to each other.

(*a*) Can you get all the coins showing heads in just two moves?
(*b*) Can you make them show all tails in any number of moves?

## Triangle of Coins

Start with a triangle of ten coins pointing upward, as shown. Can you move three coins only and make the triangle point downward?

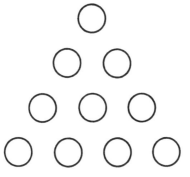

## Five-Coin Trick

Take five coins, all the same kind—say all dimes. Can you place them so that each coin touches the other four?

## Five-Coin Puzzle

Can you shift the coins shown on this board so that the penny and the half-dollar on the left swap places?

## Coin Changeovers

Place three pennies and three nickels as shown here. Can you make the pennies and nickels change places? You may move only one coin at a time. Move it directly to an empty place, or jump it over another coin to an empty space. You can move or jump up and down or across but not diagonally.

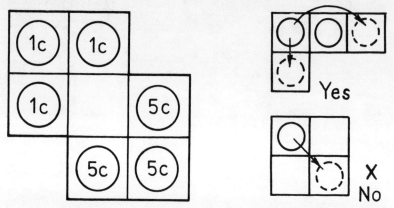

Now try this puzzle:

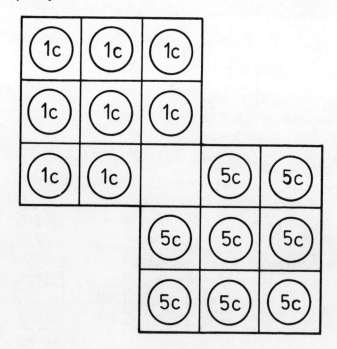

## Mission Impossible?

Two secret agents, 005 and 007, are each trying to get this top scientist out of Slobodia. The only way out is across the Red River Danube. Agent 005's man is Dr. Fünf and 007's man is Dr. Sieben. None of them can swim. A rowing boat awaits them, hidden on the Slobodian side. It carries only two people at a time. Neither scientist dare be alone with the other agent unless his own agent is also with them. Nor can the two scientists be left alone together, in case they swap top secrets. It's a case of two's not allowed, three's company!

For instance, Dr. Fünf cannot row across the river alone with the other agent, 007, or be alone with him on either river bank; but he can be on either bank when *both* agents are with him. How did the agents row the scientists across the river?

HINT: Five crossings from bank to bank should complete the mission.

## Railroad Switch

The driver of a shunting engine has a problem: to switch over the black and white cars on the triangle-shaped siding shown here. That is, he must shunt the white car from the branch $AC$ to the branch $BC$ and the black car from $BC$ to $AC$. The siding beyond is only big enough to take the engine or one car. That's all. The engine can go from $A$ to $B$, back up past $C$, and then forward along $AC$. But when it does so, it will end up facing the other way along the tracks $AB$. The driver isn't bothered about which way his engine faces. Can you switch the cars in six moves? Each coupling and uncoupling counts as one move. Remember, the driver can couple up both cars to the engine and then uncouple just one of them.

To solve the shunting problem, draw a large map of the railroad and use coins on it.

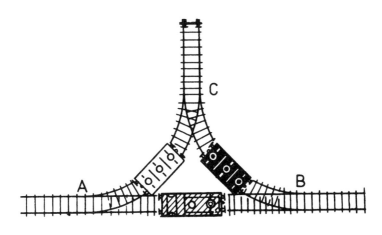

## Restacking Coins

There used to be a toy called the Tower of Hanoi. It was in the form of 64 wooden rings of graded sizes stacked on one of three pegs—largest at the bottom, smallest at the top. The rings had to be restacked in the same order on a different peg by moving them one at a time. A story goes that this problem was sent to Buddhist monks. Working at a move a second, they would have needed some 585 billion years to finish it!

Here is a new—and shorter!—version of this old puzzle. Place three saucers or table mats in a line. In the first saucer on the left stack a quarter, a penny, and a dime; the quarter must be at the bottom, the penny on top, as shown. Restack the coins in exactly the same way in the far right saucer. You must follow these rules: Move only one coin at a time, from one saucer to another. *Always* put a smaller coin on a larger coin. *Never* put a larger coin on top of a smaller one. Use all three saucers when moving the coins. You can move to and fro.

## River Crossing

A platoon of soldiers must cross a river. The bridge is down, the river wide. Suddenly the platoon's officer spots two boys playing in a tiny rowboat. The boat only holds two boys or one soldier—*not* a boy and a soldier, for instance. All the same, the platoon succeeds in crossing the river in the boat. How? Work it out with matches and a matchbox on the table across a make-believe river.

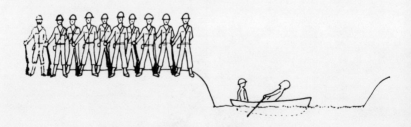

## Collision Course?

Two trains have met head on on a single track in the desert. A black engine (*B*) and car on the left; a white engine (*W*) and car on the right. There is a short switch just large enough for one engine or one car at a time. Using the switch, the engines and cars can be shunted so they can pass each other. How many times will the drivers have to back or reverse their engines? Count each reversal as a move. A car cannot be linked to the *front* of an engine.

# 6. Reasoning and Logical Problems

In this section I've included some novel thinking exercises with blocks. And there is also a selection from several types of IQ tests that are visual and mathematical in nature. The section continues with a sprinkling of some of the better known (and lesser known) logical puzzles that call for strict reasoning. I have concluded the section with some unusual logical puzzles not often seen in puzzle books.

### Thinking Blocks

The following are problems about placing six rectangular blocks so that they touch only so many other blocks. They were originally included in a book by Edward de Bono, *Five-Day Course in Thinking* (New York: Basic Books, 1967). He used the problems as a cunning thinking exercise. You may find you simply cast your six blocks on the table in random fashion and hope for the best. Of course, you've still got to check the pattern of blocks you get this way. Or you can adopt a less higgledy-piggledy approach and carefully build up a pattern, block by block. One way is as good as the other. Any method will serve just so long as it gives you the correct answer.

There is a simple lesson to be learned from this puzzle. We often cannot solve a problem because we get a thinking block. We get blocked in our thinking, or rather, in *one* way of thinking. So the lesson is this: If one way of thinking proves unhelpful, try another. Often the more ridiculous the new way of thinking seems, the better it may be.

One other tip: Don't discard ideas that didn't work. To know that a certain pattern of blocks doesn't give the right answer is in itself useful. The trick is to *remember* all these "blind alleys" so that you don't try them repeatedly and thus lose patience. Matchboxes make good homely blocks.

**A.** Place six blocks so that each touches only two other blocks. They must touch flush along their sides, you cannot have the point of one block "digging into" another. To help you solve this puzzle, you can copy each pattern you form and jot down the number of touching blocks on each block, as shown here. This pattern won't do because two of the blocks touch *three* others.

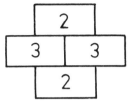

B. Place the six blocks so that each touches only three other blocks.
C. Place the blocks so that each touches four others.
D. Place them so that each touches five others.

## Martian Orders!
On Mars young Martians have to line up at school in order, according to two rules. First, girls come before boys. Second, where two girls come next to each other, the taller girl goes first; and the same goes for two boys together in a line. Zane is a Martian boy who is the same height as Thalia (a girl), but he is taller than his friend Xeron (another boy). (*a*) How do the three line up, from left to right? (*b*) They are joined by Thalia's friend Sheree (another girl), who is taller than she is. Now how do they line up?

## What Shape Next?
Here are two picture puzzles of the kind you see in intelligence tests. Follow the pattern of shapes in each from left to right. Then work out which of the lettered shapes best fits onto the end of the line.

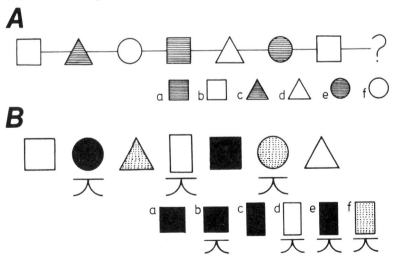

## IQ Puzzle
Another IQ-style puzzle. Look at the four numbered shapes and say which one best fits the space in the bottom right-hand corner of the picture.

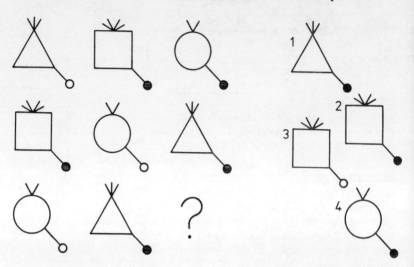

## Odd Shape Out
In each of the sets of shapes shown here one of them (1, 2, 3, or 4) is the odd shape out: It is different in some way from the other three shapes. Can you pick it out in each set?

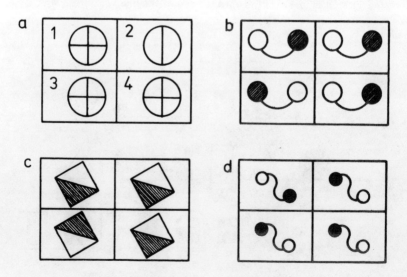

## The Same Shape
Which of the shapes shown here is the same as the boxed one on the left?

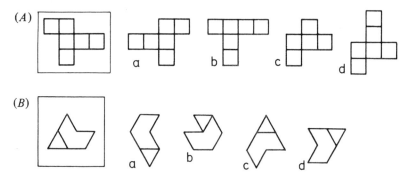

## Next Shape, Please
Can you say which is the next shape in the pattern?

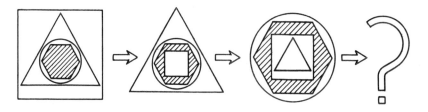

## The Apt House
Yet another of those IQ-like tests, or rather "aptitude" tests. Look at the numbered houses. Only one can be made by folding the plan. Which house is it?

If you can see this easily in your mind's eye, then the testers say you should be good at being an engineer. Or maybe you know your *apt* (apartment) houses!

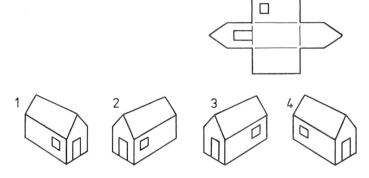

## Who Is Telling the Truth?
The judge listened quizzically to the four well-known crooks. "You're lying your heads off," he declared. "Still you'll look better that way!" The officer laughed dead on cue and then said: "I happen to know *one* of this lot *is* telling the truth." The judge snapped: "Well, what've you got to say in defense, you lot?"

Al said: "One of us is lying!"
Bob: "No, I tell you, two of us is lying."
Con: "Look here, three of us is lying."
Don: "Nope, not true! Four of us is telling the truth."
So who *is* telling the truth? The officer was quite right.

## The Colored Chemicals Puzzle
Mr. Mad the chemist has six big bottles of colored liquids. There is one of each of these rainbow colors: red, orange, yellow, green, blue, and violet. Mr. Mad knows that some of the bottles contain poison. But he can't remember which! However, he *can* remember the facts from which you should be able to work out which colored bottles contain poison.

In each of the following pairs of bottles one is poisonous, the other is not: the violet and the green bottle; the red and the yellow one; the blue and the orange one. And he also remembers that in each of the following pairs of bottles there is one that contains a nonpoisonous liquid: the violet and the yellow one; the red and the orange one; and the green and the blue one.

"And, I nearly forgot," adds Mr. Mad. "The red bottle has a nonpoisonous liquid in it."

Which bottles have poison in them?

## Mr. Black, Mr. Gray, and Mr. White
Three men met on the street—Mr. Black, Mr. Gray, and Mr. White. "Do you know," asked Mr. Black, "that between us we are wearing black, gray, and white? Yet not one of us is wearing the color of his name!" "Why, that's right," said the man in white. Can you say who was wearing which color?

## Hairdresser or Shop Assistant?
Amy, Babs, and Carol are either hairdressers or shop assistants. Amy and Babs do the same job. Amy and Carol do different jobs. If Carol is a shop assistant, then so is Babs. Who does which job?

## The Zookeeper's Puzzle
The Zookeeper wants to take two out of a possible three chimps to a TV studio. The two male chimps are Art and Bic; the third, a female, is called Cora. He daren't leave Art and Bic behind because they fight. And he cannot take both with him either. But Cora doesn't get on with Bic. So who can he take?

## Who's Guilty?
Alf, Bert and Cash are the suspects in a robbery case. Their trial shows up the following facts: Either Cash is innocent or Bert is guilty. If Bert is guilty, then Cash is innocent. Alf and Cash never work together and Alf never does a job on his own. Also, if Bert is guilty, so is Alf. Who is guilty?

## Who's in the Play?
Alice won't take part in the Buskin Players annual (amateur) play if Betty is in it! But Charles will only play if Alice *is* in it. The poor producer insists that *one* of the girls is in the play. Two people are needed. Who is in the play?

## Tea, Coffee, or Malted Milk?
The professor had enjoyed his usual after-lunch beverage so much he thought he'd have another. But he could not for the life of him remember what he had drunk. So he called the waiter over. And this is what he said to him: "Now, if this was coffee, I want tea, and if this was tea, bring me a malted milk. But if this was malted milk, bring me a coffee."

The waiter, who was logically minded, then brought him coffee. Can you say what drink—tea, coffee, or malted milk—the waiter had originally served the professor?

## Soda or Milkshake?
Three friends—Alan, Bet, and Cis—often go to the same soda fountain. Each either orders a soda or milkshake. The soda jerk notices: (*a*) when Alan chooses a soda, Bet has a milkshake; (*b*) either Alan or Cis has a soda, but never both; and (*c*) Bet and Cis never both have a milkshake. There are only two possible orders they can make. What are they?

HINT:   Since this is a hard one, we'll tell you that only Bet has a choice of drinks.

## Newton's Kittens
Isaac Newton, as you probably know, was one of the cleverest men the world has ever known. He was the great scientist and mathematician who solved the riddle of gravity, of why things fall to the ground. Well, Newton had a cat and she used to come and go into his house near Cambridge, England, through a large hole bored in the bottom of his kitchen door. One day the cat had three kittens. And so Newton had three small holes bored in the door for them.
    Why do you think this was funny?

## March Hare's Party

The March Hare was giving a party. His young guests had to get to their rightful tables—1, 2, 3, or 4—by one of the four paths shown in the picture. As you see, he wouldn't let any boys go along one path, which would later fork into two paths. Al wanted to have tea on an island with Barbra, but he refused to have tea with Silvie or Don. Sylvie said she wouldn't have tea near water. On top of this, Gary just *had* to roller-skate over one of the bridges. To make matters worse, Don and Gary wouldn't have tea with each other. By the way, only the boys could row, and the single boat was only big enough for one child and could only travel to table 4. Where did each guest have tea?

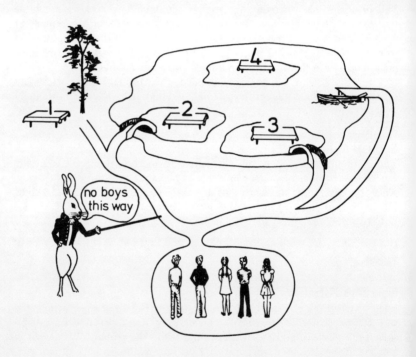

## Marriage Mix-up
The absent-minded professor had just been to a party. His wife naturally wanted to know who was there. "Usual crowd," he replied. "And some new faces. Ted, Pete, and Charlie. And their wives—Barbra, Sue, and Nicola. Can't remember who's married to whom. Anyway, each couple has one child: They're called Ruth, Wendy, and Dick. Told me all about them. Barbra said her child was playing Annie in *Annie Get Your Gun,* the school play. Pete told me his child was playing Ophelia. I do remember Ted pointing out that his daughter was not Wendy. And Charlie's wife is not Sue. I suppose we can work out the marriage partners from that."

See if you can work out who is married to whom and who their children are.

## Who Does Which Job?
There are three men—Orville, Virgil, and Homer. Each has two jobs. The jobs are: private eye, racing driver, singer, jockey, bartender, and cardsharp. Try to find each man's two jobs from these facts:

(*1*) The bartender took the racing driver's girl friend to a party
(*2*) Both the racing driver and the singer like playing cards with Homer
(*3*) The jockey often had a drink with the bartender
(*4*) Virgil owes the singer a buck
(*5*) Orville beat both Virgil and the jockey at cards

## Birds and Insects
Here's an easy logical poser—or is it? Think about these statements:

>No birds are insects.
>All swallows are birds.

Which of the next sentences follows *logically* from the above two statements?

(*A*) No swallows are insects
(*B*) Some birds are not swallows
(*C*) All birds are swallows
(*D*) No insects are birds

## Wonderland Golf
The American mathematician Paul Rosenbloom specially devised this zany golf game for youngsters. He set it as a piece of mathematical research, actually. On the Wonderland Golf Course the holes are numbered 1, 2, 3, and so on up to 18. The links are laid out in a spiral, as shown, to make the shots easier! You have two special clubs. One of them holes out in one

for you! This is the Single-Shot Iron, or S iron for short. The other club even lets you skip holes! It hits your ball from any hole to the one double its number; so it hits your ball from, say, hole 1 to hole 2, hole 3 to hole 6, and hole 9 to hole 18. Call it the D (for double) iron.

PUZZLE: What is the smallest number of shots, using either iron, to get from hole 1 to hole 18? That is, what is par for the course? Strangely, it is the same as for holes 11, 13, 14, and 17! Don't suppose you can spot a pattern, can you?

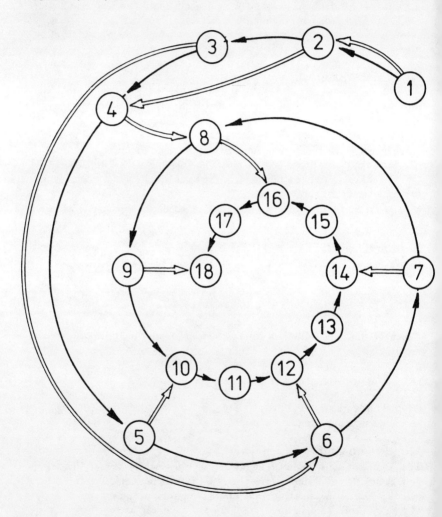

## Mad Hatter's Tea Party

The Mad Hatter had planned a special children's tea party. He had laid out the tables in the garden in the way our picture shows. He had split his guests into three sets—*G:* all girls; *B:* all boys; and *M:* boys and girls, mixed. You can see them in their sets, on the left of the picture, waiting to have tea. He told them: "Everybody in each set has to get to his table by taking the correct path through the garden. You can see which way to go by the words set in the paths."

Can you work out which of the tables—1, 2, 3, or 4—each set should get to? One of the tables remains empty. Which one?

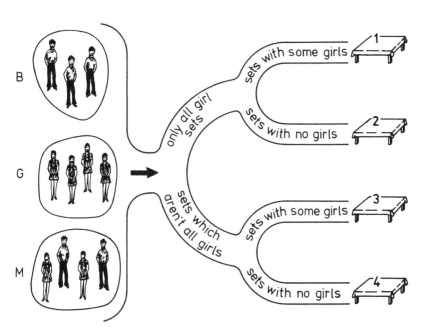

# 7. Mathematical Games

All but two of these games are mathematical; I have included two word games, "Coincidences" and "The Crossword Game," because they are so good and so popular. Do not be put off by the word *mathematical*: the games can be played without the foggiest notion of math. Indeed, "Mancala" is a very ancient African game and has been played since time immemorial by people without the slightest inkling of what we know as school math. First and foremost, the games are meant to be played to win and for fun. Any educational spin-off is purely coincidental, as they say.

### Nim
*A Game for Two*

This is one of the oldest and most enjoyable of mathematical games for two. The word *nim* probably comes from the Shakespearean word meaning to steal. Possibly it was first played in China.

Nim is played with matches or coins. In the most popular version 12 matches are placed in three rows—3 matches, 4 matches, and 5 matches, as shown.

The rules are simple. The players take turns in removing one or more matches, but they must all come from the same row. The one who takes the last match wins. (You can also play the other way: The one to take the last match loses.)

Playing a few games will soon show you how you can always win: (*a*) Your move must leave two rows with more than 1 match in a row and the same number in each; (*b*) your move leaves 1 match in one row, 2 matches in the second row, and 3 in the third; or (*c*) if you play first, on your first move you take 2 matches from the top row and after that play according to the first two winning strategies just given.

You can play Nim with any number of matches or pennies in each row, and with any number of rows. As it happens, there is a way of working out how to take the right number of matches to get into a winning position. You simply use "computer counting," or binary. This method was first given in 1901. A description of it is given in the answers section.

## Tac Tix
*A Game for Two*

Tac Tix is an exciting version of "Nim," invented by the Danish puzzlist Piet Hein. Hein is the inventor of "Hex," page 86. In Tac Tix pennies or counters are placed in a square, as shown in the picture. Players in turn remove one to four pennies from the board; they may be taken from any row or column. But they must always be adjacent pennies with no gaps between them. For example, say the first player took the two middle pennies in the top row; the other player could not take the other two pennies in that row in one move.

Tac Tix has to be played with the player taking the *last* penny losing. This is because a simple tactic makes playing the usual way uninteresting— perhaps the reason for the game's name—for it allows the second player to always win. All he has to do is play symmetrically—that is, he takes the "mirror" penny or pennies to the one(s) the first player removed. The game can also be played on a three-by-three board, and there, when playing the usual way, the first player can win by taking the center penny, a corner one, all of a central row, or all of a central column.

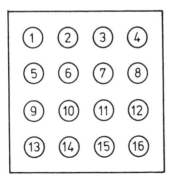

## Battleships
### A Game for Two

One of the most popular of all paper-and-pencil games, Battleships can also be a serious exercise in math! Each player has a fleet of ships, which he marks on a grid; he fires salvos at named enemy squares, and the enemy tells him if he has hit a ship or not and, if so, what kind of ship is hit. From this he tries to work out where the enemy ships lie. To sink an enemy ship, he must hit every square of that ship. First to sink the enemy's fleet wins.

Each player needs two ten-by-ten grids marked $A, B, C, \ldots, J$ along the top and $1, 2, 3, \ldots, 10$ along the left side, as shown in the picture. On one sheet he marks the positions of his fleet; the other sheet is for marking his own shots at the enemy fleet. (The second sheet represents a different area from the first; otherwise it would be possible for a player's ship and his enemy's to occupy the same spot.) The picture shows the position and size of each kind of ship.

Each player has a fleet of:  
    One battleship    (four squares each)  
    Two cruisers    (three squares each)  
    Three destroyers    (two squares each)  
    Four submarines    (one square each)

All ships except submarines, must be rectangles one square wide. No L-shaped or crooked ships are allowed; otherwise a player could not work out how ships lie from enemy reports on his salvos without tremendous difficulty. Two ships may not touch, even at corners. And a ship can have at most one side of a square on the border of the "sea." So a submarine cannot be placed in a corner.

When the two fleets have been marked in position on the grids, one of the players fires a salvo of three shots: He tells the enemy player where he wants the three shots to land; they don't have to land on adjacent squares. His enemy must then tell him how many shells fell in the sea and how many hit which types of ships, but he does not have to say which shot did what. For example, he might say, "Two in the sea and one on a destroyer." No matter what order these results were gotten. The second player now fires a salvo, and the first player tell him what happened. Each player keeps a record of his hits and misses on his chart of "enemy waters" to work out where the enemy fleet is moored. Play continues until one of the players sinks the enemy's entire fleet and announces the fact.

## Boxes
### A Game for Two
This is a game of drawing boxes on a grid of dots. It is very much like "Snake" (page 77) and can be played on the same sort of grid. Players take turns drawing a line across or down to link adjacent dots not yet linked. A player wins a box when he draws the fourth and last side of a square; he then writes his initial into the box to show he made the box. And he can then take another turn. If he's lucky, he may be able to make several boxes without his opponent having a turn. BUT after making a box he must draw one more line *immediately*. One line may make two boxes at once, but the player takes only one further turn for that line. A player does not *have* to make a box even though there may be a square with three sides drawn.

POINT ABOUT STRATEGY: Near the end of the game you usually get open "corridors" of lines, like two uprights of a "ladder." Once one player has closed off one end of the corridor (or indeed put in a rung anywhere on the ladder), the other player can make all the boxes in the corridor during his turn. The winner is the one who has made more boxes. It is best to play on a grid with an even number of dots on each side—eight by ten, say—so that there will be an odd number of boxes in the completed grid.

## Mastermind
### A Reasoning Game for Two
This game is marketed, although the principle is simple enough for you to make your own version. The basic idea is this: One player sets a problem by inserting five colored pegs, out of a possible eight colors, in a row. His pegs are then covered, and his opponent, the "mastermind," has to work out what the colors and their correct places are by forming trial rows. The problem setter indicates by the use of black and white pegs whether or

not, first, the opponent has the right colored pegs and, second, they are in the right place.

The commercial board is made of plastic and has rows of five holes with a square grid at the end of each row. The version shown here has only four holes with a two-by-two square grid at the end of each row. A sample game will serve to indicate the rules and method of play. To simplify things, we will play with four colors and white only.

Suppose the first player puts up these four pegs: green (G), blue (B), red (R), and yellow (Y). He then covers them with a little hood, or *cloche*, so that his opponent cannot see the pegs. The opponent puts in the top row: red, green, white (W), and green. As the sketch shows, he has two colors, right, but they are not in the right place; to show this, the first player puts in two white pegs. The opponent's second try is the line green, blue, black (Bk), and white. Because this row has two colors *and* two places right—the green and the blue pegs—the first player puts in two black pegs. The opponent's third row is green, blue, white, and red, which has two colors in the right place (two black pegs) and one color right (red) but in the wrong place (one white peg). The game ends when he has formed a row exactly the same as the one originally set. The opponent works out by reasoning which pegs to change. The shorter the number of rows he can solve the problem in, the better he is at reasoning. The game can also be played on paper, with colored pencils or felt-tip pens substituting for pegs.

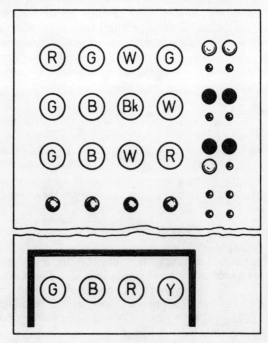

## Coincidences
### A Word Game Any Number Can Play
This is a word game rather like "Mastermind" but played with letters instead of colored pegs. One player acts as the "accountant," who thinks of a five- or six-letter word. He notes this secretly on a sheet of paper, which he keeps. He calls out the number of letters it has. The other players try to discover his word by calling out a line of the same number of letters. The accountant then tells each player how many of the letters in his line match in position those in his word. Say the accountant's word is "CENTS" and a player calls "A-A-A-E-E"; then the accountant announces "None" because although the player has gotten one letter right ($E$), neither $E$ is in the correct position. A good strategy for players is to call out all vowels, since most words contain them ($A, E, I, O, U$). If the accountant's word is discovered in fewer tries than there are letters in his word (four tries, say), then he scores nothing. He scores one point for every try over the number of letters in his word.

Here is how a sample game began. Accountant's word: SHIRT

| Player's lines | Accountant calls |
|---|---|
| AAEEE | None |
| IIOOO | None |
| III̲OO | One |
| THI̲TH | Two |
| THI̲TT | Three |
| THI̲LT̲ | Three |

To play well, one should know the letter frequences of English, as follows:

*Single letters:*    E, T, A, O, N, I, S, R, H, L, D, C, U
*Two-letter groups:*    TH, IN, ER, RE, AN, HE, AR, EN, TI, TE, AT, ON, HA
*Three-letter groups:*    THE, ING, AND, ION, ENT, FOR, TIO, ERE, HER, ATE, VER, TER, THA

## Eleusis
### A Reasoning Game for Four
Here is a game with a really novel twist, invented by a New Yorker, Robert Abbot (taken from *More Mathematical Puzzles and Diversions,* by Martin Gardner, New York: Penguin, 1961). He originally devised it as a card game, but it can also be played with paper and pencil. Its novel twist is this: Most games have rules you learn and use in order to decide your best move, but in Eleusis you play to discover the rule! The game is rather like discovering a scientific law, except that in science there is nobody to tell you if your law is the right one (or indeed if there even *is* such a law).

In Eleusis an "umpire" secretly sets the rule, which other players have to discover. There are several games based on Eleusis; ours is the paper-and-pencil version for four players—an umpire and players A, B, and C. (It could also be played by just two players.) Our game consists of four sets—one for each player. When everybody has had a go at being umpire, the game ends, and the player with the lowest number of penalty points is the winner.

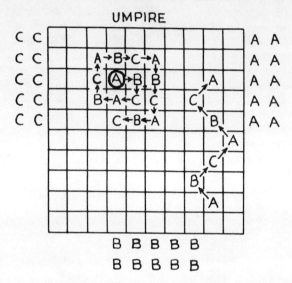

Two simple umpire's rules for the players to discover are shown in the picture of the board. On the left the rule is: The letters $A$-$B$-$C$-$A$-$B$-$C$-$A$ ... go in a spiral; the letters can move in any direction so long as their position is correct. On the right the rule is: The first letter $A$ can go anywhere; the next letters must follow a "one square up, one square to the side" rule—that is, on the diagonal. The umpire keeps a drawing of his rule to show afterward in case of dispute.

Each player arms himself with a sheet of squared paper to keep a record of both his correct moves and incorrect attempts; about ten-by-ten squares should be big enough (see picture).

Player A sits on the umpire's left, then B, then C. Each player has ten of his own letters, which he tries to correctly place on the board. Player A takes a turn by pointing to an empty square and asking if he may write a letter $A$ in it. If the umpire says yes, he puts an $A$ in that square and crosses off one of the $A$'s on his side of the sheet. If the umpire says no, the turn passes to player B, and player A cannot cross out one of his letters. When A, B, and C have each had a turn, a round is completed; the next round begins when A takes his next turn.

The idea is that the umpire should set rules that are neither too easy nor too hard for the players to discover; ideally the players should be rid of their letters in the fifteenth round. The umpire is penalized if either his rule is so easy that the players are out before the fifteenth round or if the rule is so hard that the game continues after that round. The umpire keeps a tally of his "score" (penalty) by circling numbers on the tally card shown here, one number for each round:

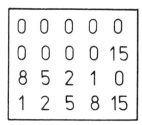

For the first nine rounds the umpire circles 0's and loses no points, since it would be impossible for the players to get rid of their letters before the tenth round. At the tenth round he is penalized 15 points; were the game to be over then, the players would have known the rules from the beginning. The penalty is progressively reduced until the fifteenth round, when it is again 0. After that round the penalty is progressively increased until the twentieth round, when it is back up to the maximum, 15.

## Ticktactoe
### A Game for Two

Ticktacktoe, or Noughts and Crosses, has to be the oldest battle of wits known to children and adults alike. The object is for one player to complete a line—horizontal, vertical, or diagonal—by himself. Any astute player will learn how to play to a draw in only a few hours' practice. The game must end in a draw unless one player makes a slip.

There are just three possible opening plays shown by the $X$'s in the picture—into a corner, the center, or a side box. The second player replies with O's. He can save himself from being trapped by one of the eight possible choices; he can mark the center. The side opening (third picture) offers traps to both players; it must be met by marking one of four cells.

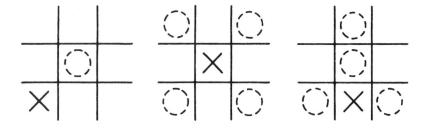

## Ticktacktoe with Coins
*A Game for Two*

A more exciting variant of "Ticktacktoe" is where you play with coins or counters that can be moved after being placed. The game was played a lot in England in the 1300s, when it was called "Three Men's Morris," the forerunner of "Nine Men's Morris," page 75. It was also popular in ancient China, Greece, and Rome.

You use six coins in all, three silver dimes, say, for one player and three pennies for the other, on a three-by-three board. You take turns placing a coin on the board until all six coins are down. By this stage either player could have won by having three of his own coins in a row—horizontal, vertical, or diagonal. If neither player has won, they continue playing by moving a single coin one square, across or down, to any empty square. Diagonal *moves* are not allowed.

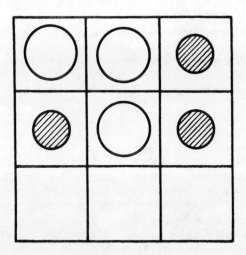

## Teeko, or Five-by-Five Ticktacktoe
*A Game for Two*

A modern variant of "Ticktacktoe with Coins" called Teeko was invented by the American magician John Scarne. It is played on a five-by-five board. Each player takes turns placing four coins. Then each moves in turn one square horizontally, vertically, or diagonally. A player wins by getting his four coins in a square pattern on four adjacent squares or four coins in a horizontal, vertical, or diagonal row.

# Nine Men's Morris
## A Game for Two

This old English game was known to Shakespeare and has been played by young and old ever since. It is really a variant of "Ticktacktoe." It used to be played in the village green on a pattern like this one scratched in the earth. Nowadays it is played on a board. You can copy this diagram onto a sheet of stiff paper.

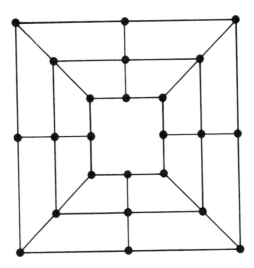

You need eighteen "men," nine black and nine white. Tiddledywink discs in any two colors will do well. Each player has his own color men. Each player in turn puts one man down on one of the 24 dots. He tries to form a three-in-a-row, or *mill* as it is called. A mill must be in a straight line; it cannot bend around a corner. This is a mill:

This is not a mill:

A player who has put down a mill can, on that move, remove an opposing man, providing it is not already part of a mill. The loser of the game is the player who loses all his men first. The shape I show here is the commonest form used; but other shapes are possible. In fact, why not invent one of your own?

## Peggity
### A Game for Two

This is an ancient game of position played on squared paper. It is also known as Peg Five and Spoil Five. Thousands of years ago in China the game went by the name of Go-Moku. It is like the famous Japanese game Go, except that Peggity does not involve capture of enemy pieces and so can be played on paper with pencil.

The board is 19-by-19 squares big; the pieces, $X$'s and $O$'s, are played one at a time into any of the sqaures (not on the corners, as in Go.) $X$ moves first. The aim is to be first to get a straight line of exactly five adjacent $X$'s or $O$'s all in a line along a row, column, or either diagonal. Each player has as many $X$'s and $O$'s as he needs.

A player with four $X$'s (or $O$'s) in a line—known as an open-four—must win on his next move because his opponent cannot block both ends in one move. But when he has another $X$ one space away at the end of the line (see picture), all the opponent has to do is play an $O$ at the other end of the open-four. If the open-four player plays into the space, he gets a line with six $X$'s in a row; this does *not* qualify as a line of five symbols. After forming an open-three—three $O$'s or three $X$'s in a line—it is usual to call "three." This is because it can become an open-four on the next turn and thus a potential winning position. Calling three avoids the likelihood of a player losing by an oversight, which is fun for neither. A joined pair of lines of three $X$'s (or three $O$'s) is called a double-three (see picture).

## Snake
*A Game for Two*
This game is played on a five-by-six board of dots, like this one. Players take turns at joining two dots by a line to make one long snake. No diagonal lines are allowed. You cannot leave any breaks in the snake. Each player adds to the snake at either end; a player can only add to his opponent's segment, not to his own. The first to make the snake *close* on itself loses. Here is an actual game. In it straight lines began and lost.

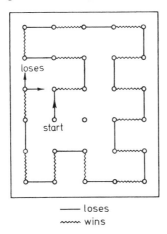

───── loses
∿∿∿∿ wins

## Daisy
*A Game for Two*
The two players take turns to pluck from the daisy either one petal or two adjacent petals. The player taking the last petal is the winner. This is a game invented by the great puzzlist Sam Loyd.

Make a daisy with 13 petals out of matches, like this. On a postcard mark little circles where the petals (matches) grow from. You need to know whether you have left a space between petals or whether petals are next to each other. The second player can always win—if he knows how. See the answer section for this winning strategy.

Remember you cannot take two petals if there is a space between them. That's why we recommend marking the petals' positions.

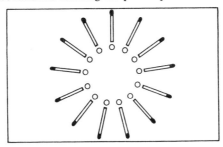

## Sipu
### A Game for Two

Sipu is an old folk game from the Sudan. It is like "Ticktacktoe," but it is different in two ways: It hasn't got an obvious strategy for not losing, and you can move your $X$'s and $O$'s as counters after they've been put down. It is played on a square board with any odd number of squares along each side; the odd number ensures that there is a center square. (Actually, Sipu is the name of the game played on a five-by-five board. The three-by-three board game, described here, is called Safragat.) You need counters or pebbles, known as "dogs," of two colors, or two kinds of coins (pennies and dimes, say). We'll call them Blacks and Whites. In order to see how play goes, we'll start with a three-by-three board, for which you need four Blacks and four Whites.

Play goes in two stages: First placing the counters and then making the moves and taking the opponent's counters. The best way to place your counters will become clear after playing a few games. To see who starts the placing, toss for it or conceal a different coin in each fist and let your opponent guess. Say Black begins the placing. Then you place the counters in turn—a Black, then a White, and so on—until the board is filled, leaving the center square empty. Let's say you have filled the board like this:

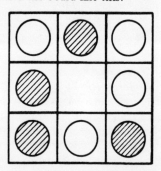

You are ready to begin moving counters. Toss to see who moves first. Let's say it is White's first move. Counters are moved either up or down or side to side, but *not* diagonally. They can only be moved into an empty square.

**(1)** White (we'll say) moves into the empty center square.

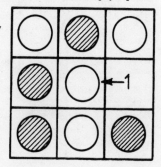

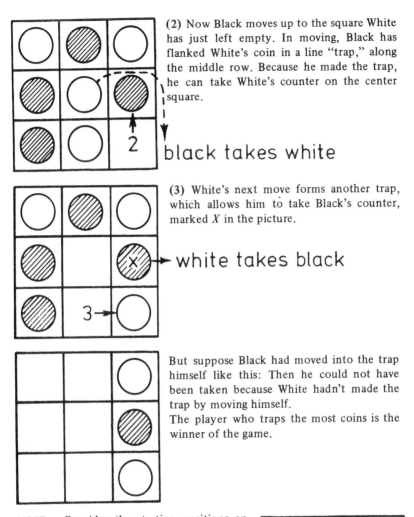

(2) Now Black moves up to the square White has just left empty. In moving, Black has flanked White's coin in a line "trap," along the middle row. Because he made the trap, he can take White's counter on the center square.

black takes white

(3) White's next move forms another trap, which allows him to take Black's counter, marked $X$ in the picture.

white takes black

But suppose Black had moved into the trap himself like this: Then he could not have been taken because White hadn't made the trap by moving himself.

The player who traps the most coins is the winner of the game.

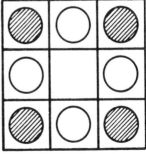

NOTE: Consider the starting positions on this board. White can move the center counter in the bottom row up, but Black still would have trapped him on the second move. In this starting position Black cannot move at all. Clearly, a position to be avoided!

Longer-playing and harder games can be played on a five-by-five board (with 12 players apiece) or even a seven-by-seven board (with 24 players apiece). The same rules apply.

## Mancala
### A Game for Two

Mancala is an old African game now coming into fashion in America and Europe. Two players—we'll call them Al and Fey—sit on either side of a board, which is usually about a foot long with six hollows on each side. (The hollows can also be scratched in the ground.) At the start of the game each hollow is filled with four stones, balls, beads, or pebbles. The aim is for one player to capture all the others' stones; these are the loot.

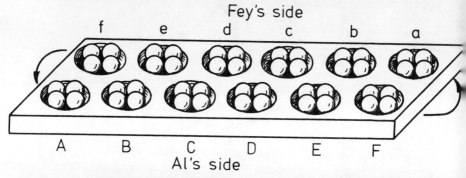

A player moves by taking all the stones out of one of the hollows on his own side and dealing them out in order, *counterclockwise* around the board, 1 stone into each hollow. The players move in turn. We've lettered the hollows simply to show how the game goes. Al's are *A, B, C, D, E,* and *F*; Fey's are *a, b, c, d, e,* and *f*. Say Al empties hollow *E*: He deals 1 ball each in *F, a, b,* and *c*. Then Fey empties, say, *b* (which now holds 5 stones). She deals them out into hollows *c, d, e, f,* and *A*. The board then looks like this:

$$\begin{matrix} f & e & d & c & b & a \\ 5 & 5 & 5 & 6 & 0 & 5 \\ 5 & 4 & 4 & 4 & 0 & 5 \\ A & B & C & D & E & F \end{matrix}$$

How does a player take loot? By placing the last stone in his opponent's last hollow (*F* or *f*) so that there are 2 or 3 stones there. Here are three possible moves to illustrate the point.

1. This is the setup. It is Al's move.

$$\begin{matrix} f & e & d & c & b & a \\ 1 & 2 & 2 & 3 & 1 & 2 \\ 0 & 0 & 0 & 0 & 0 & 6 \\ A & B & C & D & E & F \end{matrix}$$

Al moves all 6 stones from $F$ (his only move), giving:

$$\begin{matrix} f & e & d & c & b & a \\ 2 & 3 & 3 & 4 & 2 & 3 \\ 0 & 0 & 0 & 0 & 0 & 0 \\ A & B & C & D & E & F \end{matrix}$$

Al's last stone went into $f$, which now has 2 stones. He takes these, together with the 3 stones in hollow $e$ and the 3 in $d$. His loot does not skip back over $c$ to collect $b$ and $a$. He wins a total of 8 stones.

2. Here's another setup. Again it is Al's move.

$$\begin{matrix} f & e & d & c & b & a \\ 0 & 2 & 3 & 0 & 3 & 1 \\ 1 & 0 & 0 & 0 & 7 & 8 \\ A & B & C & D & E & F \end{matrix}$$

Moving from $F$, Al would win no loot, since his last stone would go in $B$, on his own side of the board. Moving from $E$, he also would win nothing; his last stone would go in $f$, which it must do to collect the loot, but does not result in 2 or 3 stones in that hollow.

3. Empty hollows aren't necessarily safe. Here all but one of the hollows on Al's side are empty. But the game still goes on.

$$\begin{matrix} f & e & d & c & b & a \\ 18 & 0 & 0 & 0 & 1 & 0 \\ 0 & 1 & 0 & 0 & 0 & 0 \\ A & B & C & D & E & F \end{matrix}$$

Fey deals from hollow $f$:

$$\begin{matrix} f & e & d & c & b & a \\ 0 & 1 & 1 & 1 & 2 & 2 \\ 2 & 3 & 2 & 2 & 2 & 2 \\ A & B & C & D & E & F \end{matrix}$$

The last stone goes in $a$. She takes all the stones from Al's side. Why are there now no stones in $f$? Why didn't Fey put a stone in $f$? Because she took 12 or more stones out of it. When you take 12 or more stones out of a hollow, you skip that hollow when you come to it; thus the twelfth stone in Fey's hand goes in the next hollow. The game ends when the players agree there are not enough stones left to form loot or when a player cannot make a move.

## The Crossword Game
### A Game for Two to Five

This is such a popular game that though it is not a mathematical game, I have put it in. Each player has his own five-by-five grid. After it is decided who plays first, the first player calls out a letter and writes it in one of the 25 squares on his grid. Each of the other players writes the same letter in some square on his own grid. The next player then calls out a letter—the same or another one—which the other players enter on their grids. Each player must write his letter before the next letter is called. The aim is to form words reading across or down. If you cannot form a word, you can call an "unuseful" letter such as $Z$ or $Q$ so that nobody else is likely to be able to use it. Only words from an agreed-upon dictionary count, not proper names or slang words. When 25 letters have been called out and each player's grid is filled, scoring begins.

Two-letter words score 2; three-letter words score 3; four-letter, 4; and five-letter, 6 (an extra point). Totals for across and down are added together to get the final score. Highest score wins. Two words in the same row, or column, may not share the same letters. For example, the letters $H$-$E$-$A$-$R$-$T$ score 6 points; whereas $H$-$E$ scores 2 and $A$-$R$-$T$ scores 3 points, totaling only 5 points. You could not add the two together to make 11. Also you could not score 2 for $H$-$E$ as well as 4 for $H$-$E$-$A$-$R$.

Here is a complete grid with its scores and a grand total of 36.

| C | H | E | S | S | 6 (chess) |
|---|---|---|---|---|-----------|
| H | E | A | R | T | 6 (heart) |
| V | I | N | E | R | 4 (vine)  |
| T | O | T | O | Y | 4 (toto)  |
| O | P | O | L | O | 4 (polo)  |
| 2 | 2 | 3 | 2 | 3 | (36)      |
|(to)|(he)|(ant)|(re)|(try)| |

# The Cop and the Robber
*A Game for Two*

Here is a single-board game for two. You can play on the city plan shown here or draw a larger version for yourself.

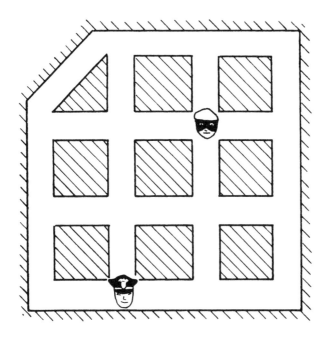

You need two coins, one for the cop, the other for the robber. Start with each coin on its picture. The rules are simple: The cop always moves first. After that, the players take turns to move. You move a coin one block only, left or right, up or down—that is, from one corner to the next. The aim is for the cop to catch the robber, which is done by the cop landing on the robber on his move. To make the game interesting, the cop must catch the robber in 20 moves, or he loses.

HINT: There *is* a way for the cop to nab the robber. The secret lies in the top left corner of the city plan.

## Sprouts
*A Game for Two*

Sprouts is one of the best of the really new paper-and-pencil games. A Cambridge (England) mathematician invented it in the 1960s. Its name comes from the shapes you end up with. It is a game of *topology,* a branch of mathematics which is very briefly explained in "The Bridges of Königsberg" (page 25). Topology is the geometry of floppy rulers, wiggly lines, and stretchy sheets of paper!

This is how Sprouts is played. On a clean sheet of paper begin by drawing three or four spots. We'll work with three spots.

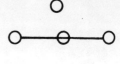

Each of the two players takes turns at joining the spots with lines, which can be as wiggly as you like. You must put a new spot somewhere along that line.

No lines may cross.

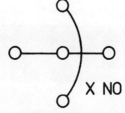

You can draw a line from a spot back onto itself to make a loop—with, of course, a new spot on it.

a loop

A spot is "dead" when it has three lines leading to it; no more lines can connect to it. To show it is dead, put a stroke through it or shade it in.

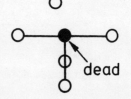

dead

The winner is the one who draws the last line. A good way to win is to trap "live" spots inside loops so that your opponent cannot use them.

Mathematicians have worked out how many moves the game can go on for: The number lies between twice and 3 times the number of spots you start with. Starting with 3 spots, the game can go on for between 6 and 9 moves; with 4 starting spots, between 8 and 12 moves. And so on. But nobody has *proved* this yet! Here is a sample game. In it player A wins in 7 moves.

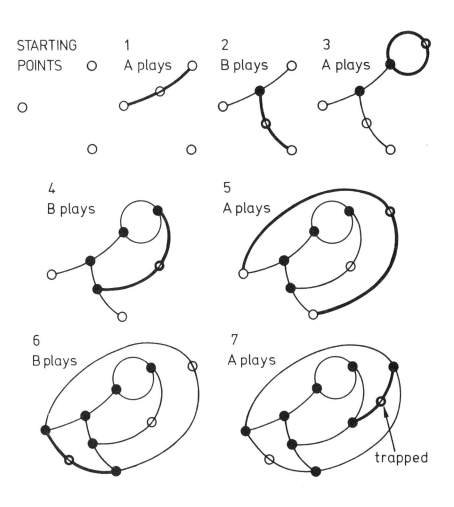

## Morra
### A Game for Two
This very old finger game comes from Italy. One player is called Morra. On a given signal—a nod of the head, for example—both players put up either one or two fingers both at the same time. The rules are in summary form:

> Both players show same number of fingers: Morra wins two pennies
> Morra two fingers, opponent one finger: Morra loses one penny
> Morra one finger, opponent two fingers: Morra loses three pennies

See if you can find a strategy that cuts Morra's losses or even lets him win. The best strategy is given in the answer section.

## Hex
### A Game for Two
Only recently invented in Denmark, this is a marvelous game, which is also called Black and White. It seems absurdly simple but is open to very cunning play or *strategy*, as it is called. The game is played on a diamond-shaped board made up of either hexagons, hence the name, or triangles.

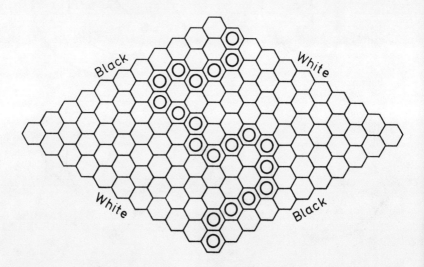

The board usually has 11 hexagons (or triangles) on each side. Two opposite sides of the diamond are Black's side; the other two are White's. The hexagons at the corner of the board belong to either player. The players take turns marking hexagons. White marks his point with a circle, Black with a heavy blob. The aim is to connect opposite sides of the board with an unbroken line of dots or blobs. (On the triangle paper, two dots are adjacent if there is a linking line between them.)

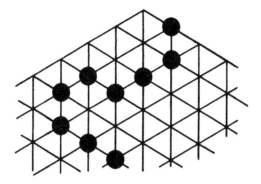

The first player to make an unbroken line is the winner. Two lines cannot cross, so there can never be a draw. Mathematicians have proved that the first player can always win, but they don't say how he is to do so! You can buy special printed paper with a grid of hexagons or triangles printed on it. If you draw up your own board, as here, do so in ink and play in pencil lightly so you can rub out the circles after each game.

To learn some of the strategies of Hex, play a game on a 2-by-2 board with just four hexagons. The player who makes the first move obviously wins. On a 3-by-3 board the first player wins by making his first move in the center of the board. This is because the first player has a double play on both sides of his opening cell, so his opponent has no way to keep him from winning in the next two moves. On a 4-by-4 board (see picture)

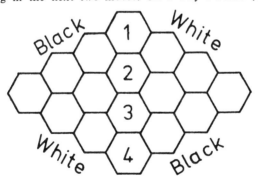

things are more complicated. The first player will win if he plays in one of the four numbered cells, but if he plays in any other cell, he can always be defeated. For an 11-by-11 board, as shown, the play is far too complicated to be analyzed.

**87**

# ANSWERS

## 1. Flat and Solid Shapes

*Real Estate!*
The combined length of the two shorter sides of the triangular plot come to the same as the long side: 230 + 270 = 500. The plot is merely a straight line and covers no land!

*Three-Piece Pie*
Find the middle of the crustless triangle and make cuts from each corner of the pie to the middle. Otherwise, you could measure the angle of the slice and divide it by 3.

*How Many Rectangles?*
Nine.

*Squaring Up*
Seven squares.

*Triangle Tripling*
Counting the little triangles in each corner gives three lots of 13 plus the big black triangle in the middle, making 40 in all. So you have 1, 4, 13, 40 triangles. Note the pattern of differences between adjacent numbers (4 − 1 = 3, 13 − 4 = 9, 40 − 13 = 27). Each difference is 3 times the previous one—as you would expect from triangles!

*The Four Shrubs*
Plant three of the shrubs at corners of an equilateral triangle; plant the fourth shrub on top of a little hillock in the middle of the triangle so that all four shrubs are at the corners of a *tetrahedron* (triangular pyramid). See answer to "Triangle Quartet" (page 110).

*Triangle Teaser*
a. 13,   b. 27.

*Triangle Trickery*
Fold the paper over as shown here. The folded flap (its underside showing uppermost) will conceal a third of the triangle's face-up area still flat on the table. It is now only two thirds of the original triangle's face-up area. So you have one third taken from two thirds, leaving one third of the original area. It therefore shows one third of the original triangle.

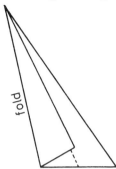

*Fold 'n Cut*
Two holes.

*Four-Square Dance*
Seven different ways.

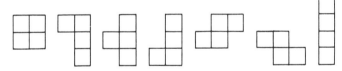

*Net for a Cube*
There are eleven nets that form a cube. The first six are, perhaps, fairly obvious; the other five you might not have thought of.

*Stamp Stumper*
The other ways are: 3 stamps joined side to side in a row, and three other **L**-shapes, like this ⊓, ⌈, and ⌋.

*The Four Oaks*

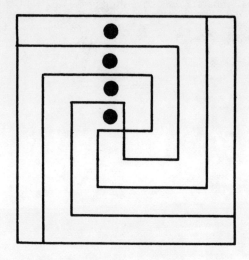

*Box the Dots*

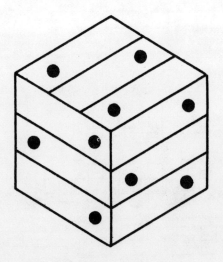

*Cake Cutting*
16 pieces. The rule is shown by the table. One cut plainly gives two pieces. For two cuts you add 2 to that number to make 4. For three cuts you add 3 to the 4 to make 7. For four cuts you add 4 to the 7 to make 11. When you draw the lines, the third line will cut two lines already drawn; the fourth line cuts three lines already drawn.

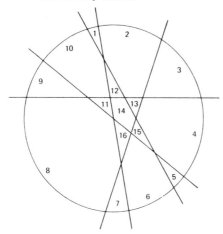

This table shows the number of pieces made by various numbers of cuts.

| no. of cuts | no. of pieces |
|---|---|
| 0 | 1 |
| 1 | 2 |
| 2 | 4 |
| 3 | 7 |
| 4 | 11 |
| 5 | 16 |

*Four-Town Turnpike*
The shortest network is made up of two diagonal turnpikes; each is $\sqrt{2}$ times 10 miles long, or 14.14 miles. So the total length of turnpike is 28.28 miles, or about 28.3 miles. The $\sqrt{2}$ comes from Pythagoras's theorem. It shows that with a right-angled triangle (two of which are produced by bisecting a square with a diagonal), where the shorter sides (the sides of the square) are each 1 unit long, the long side opposite the right angle (the diagonal) is $\sqrt{2}$ units long, or the square root of 2 units.

### Obstinate Rectangles
In a six-by-seven rectangle, the diagonal cuts 12 squares. Rule: Add the length to the width and subtract 1.

### One Over the Eight
1 + 8 jigs = 81. A jig must have 10 squares in it: $1 + (8 \times 10) = 81$.

### Inside-out Collar
To follow these instructions, it's best to label the corner of the tube $a$, $b$, $c$, and $d$ around the top edge and $A$, $B$, $C$, and $D$ around the bottom edge, as shown in picture 1.

As shown in picture 2, push corner $c$ down into the tube to meet corner $A$; this will pull the corners $b$ and $d$ together. As shown in black in picture 3, the square $CcdD$ is already inside out, as is the square $CcbB$. The triangular part with the edge $Aa$ still has to be turned inside out. This is done by pulling the corners $B$ and $D$ apart and pushing the peak ($a$) of the triangle down to meet corner $c$—like pushing someone's head ($a$) down between their knees ($B$ and $D$). Pull the corners $b$ and $d$ outward to turn the "beak" $BCD$ inside out (picture 3). You will now find that as the tube unfolds, it is inside out. The trick needs practice to perform it well: The secret is to do it in two stages—first stage is up to picture 2; second stage is pushing the peak down "between the knees."

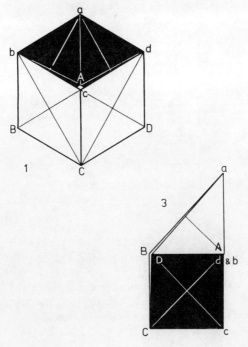
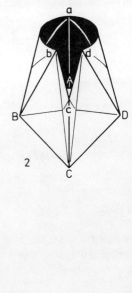

*Cocktails for Seven*

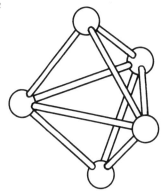

*The Carpenter's Colored Cubes*
He cut the cube into eight equal blocks, as shown.

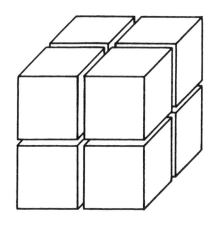

*Painted Blocks*
18 faces are painted, as shown.

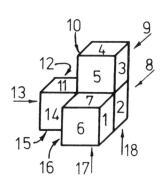

*Instant Insanity*
Take the cube marked 1 in the picture in the problem; it has three dotted faces. Place it so that two of these faces are not on any of the long sides of the rod. Next, take cube number 2 and place it so that the four different colors of it are on the long sides. Then place cube number 3 so that one of its white faces is hidden and both hatched faces are on the long sides. Place cube number 4 so that neither of the hatched faces appear on the long sides. All you have to do now is twist the cubes around the rod's axis until the solution shows up.

*The Steinhaus Cube*
Start building your cube by making this stepped shape. The rest should fit together easily.

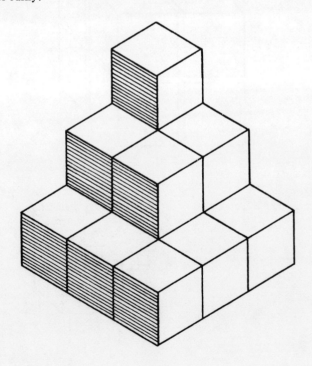

*How Large Is the Cube?*
The surface area of the cube is 6 times the area of one of its six faces. Suppose the cube has an edge $x$ inches. One of its faces has an area of $x^2$ square inches. So its total surface area is $6x^2$ square inches. But this must be equal in number to its volume or $x \times x \times x = x^3$ cubic inches. So $6x^2 = x^3$, which means $6 = x$. So the cube has a side of 6 inches.

If this reasoning is too hard to follow, go from the equation $6x^2 = x^3$ and then try $x = 1, x = 2$, and so on.

*Plato's Cubes*
The problem calls for a number which when multiplied by itself twice over gives a square number. This works with any number that is itself already a square. The smallest square (aside from 1) is 4; so the huge block might have $4 \times 4 \times 4$, or 64, cubes in it, and this would stand on a $8 \times 8$ square. The picture suggests that a side of the plaza is twice the extent of a side of the block. So this is the correct answer. The next size for the cube is $9 \times 9 \times 9 = 729$; this cube would be standing on a $27 \times 27$ square, which, according to the picture, is too large.

*The Half-full Barrel*
All they had to do was tilt the barrel on its bottom rim. Say the barrel was exactly half full. Then when the water is *just* about to pour out, the water level at the bottom of the barrel should just cover all the rim. That way half the barrel is full of water; the other half is air space.

*Cake-Tin Puzzle*
10 inches square—that is, twice the radius.

*Animal Cubes*
27 cubes in each animal. Both volumes are 27 cubic centimeters. Areas: dinosaur 90, gorilla 86.

*Spider and Fly*
Six shortest ways; each goes along only three sides. A typical way is shown by the solid lines on the cube here.

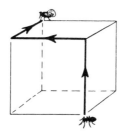

*The Sly Slant Line*
5 inches. The slant line must be the same length as the radius because it is one of the two equal diagonals in the rectangle.

## 2. Routes, Knots, and Topology

*In-to-out Fly Paths*
He can for shapes with an odd number of sides—the triangle, the pentagon, and the seven-sided shape (heptagon). As he begins inside, he has to cross an odd number of sides in order to end up outside.

*In-to-in Fly Paths*
He can for shapes with an even number of sides—the square, the hexagon, and so on.

*ABC Maze*
Out at $A$ only, because an odd number of paths (5) lead to $A$. An even number of paths lead to $B$, or to $C$; so you cannot leave by them.

*Eternal Triangle?*

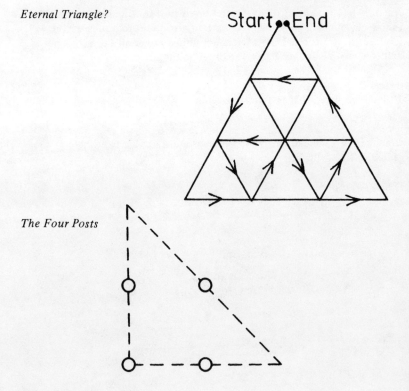

*The Four Posts*

*The Nine Trees*

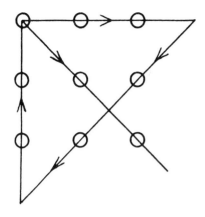

*Salesman's Round Trip*
Shortest route is *ACDBA*, which comes to 8 + 6 + 6 + 7 = 27 miles.

*Swiss Race*
Clear roadblock on road *AC*; then take route *ACDE*, which is 11 miles.

*Get Through the Mozmaze*
He can escape with 37 bites.

*Space-Station Map*
As Sam Loyd said, the more than fifty thousand readers who reported "there is no possible way" had all solved the puzzle! For that is the sentence that makes a round-trip tour of the space stations. (Actually Loyd used canals on Mars, not space stations.)

*Round-Trip Flight*
The sketch is:

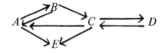

There is only one way from Albany (*A*) to Detroit (*D*) and return. El Paso is the "trap":

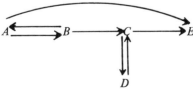

*Faces, Corners, and Edges*
Euler's rule is the number of faces ($f$) and corners ($c$) equals the number of edges ($e$) plus 2. It works for all solid figures that don't "bulge in" and don't have holes in them. It works for all the figures shown.

*Tetrahedron:* $f = 4, c = 4, e = 6$
*Octahedron:* $f = 8, c = 6, e = 12$
*Dodecahedron:* $f = 12, c = 18, e = 28$
*Icosahedron:* $f = 20, c = 12, e = 30$

*Five City Freeways*
Ten roads. Put five dots on paper and join them with lines for the roads. You'll need ten lines, probably with five crossings. By redrawing, you can reduce this number to one crossing, which is unavoidable.

*The Bickering Neighbors*

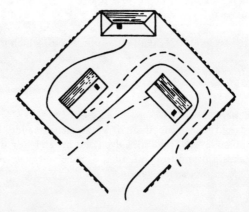

*The Bridges of Königsberg*
No, it is impossible to cross every bridge once and only once in a single stroll. Such a stroll, if you were drawing, you could call a *one-stroker*. Euler discovered that there's a simple rule for telling whether or not a route is a one-stroker. First draw the *network*, as he did for Königsberg; this cuts out all fiddly details that don't matter. Then count the number of roads (lines) leading into each dot. Call the dots *odd* if they have an odd number of lines leading into them, or *even* if the number is even. Euler found this rule: A network with all even dots or with just two odd dots is a one-stroker; it can be traced in one motion without lifting the pencil or going over the same line twice. Networks with any other number of odd dots are definitely *not* one-strokers. If you are showing somebody how to trace a network with two odd dots, be sure to begin at one of the odd dots.

*Euler's Bridges*
*Odd number of bridges rule:* The number of times you touch the north bank (call it $N$) equals half of 1 more than the number of bridges ($b$). Or in letters: $N = (b + 1)/2$.
*Even number of bridges rule:* Here the number of "norths" is 1 more than half the number of bridges, or $N = 1 + b/2$.
*Mathematical note:* To arrive at these formulas, you have to guess and juggle a bit. Note, however, that in the "odd" formula you halve $(b + 1)$; in the "even" formula, on the other hand, you only halve $b$, which is obviously possible because $b$ is even.

*Möbius Band*
Cut it down the middle and it falls into just *one* single band—an ordinary twisted collar. By cutting, you have added an edge and a face. Cut a third in, and you get one twisted collar and one smaller Möbius band linked to it.

*Double Möbius Band*
Open it out and you see it is actually one large band.

*Release the Prisoners*
You can get away from your friend by slipping the rope loop over one of your hands and then back under one of the loops around his wrist, as shown here.

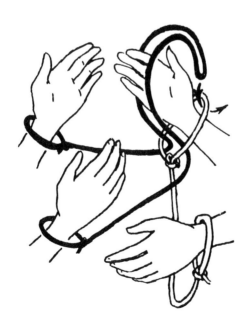

*Three-Ring Rope Trick*

*Wedding Knots*
Here are three different ways of joining the straws to make one single closed loop: There are many more ways.

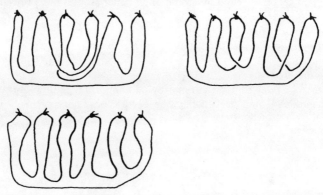

*Tied in Knots?*
Ropes *a, d, e,* and *g* will tie in a knot. But knot *h,* of course, does not tie in a knot, which is why magicians use it!

*The Bridges of Paris*
Yes, it was possible to take such a stroll. The network shown here has two odd points, so it must be possible according to Euler's rule. See "The Bridges of Königsberg."

*Tour of the Castle*
These problems are obviously related to "Euler's Bridges." But to find a general rule is not so obvious! To find one, you have first to draw networks as Euler did in "The Bridges of Königsberg." The answers are: (*1a*) yes, (*1b*) no, (*1c*) yes, (*2a*) no, (*2b*) yes, (*2c*) no, and (*2d*) yes.

*The Cuban Gunrunners Problem*
Blow up bridges *a*, *c*, and *d*.

## 3. Vanishing-Line and Vanishing-Square Puzzles

*Mr. Mad and the Mandarins*
The oranges on the plate of the absent child "vanished." But, actually, each remaining plate of mandarins has gotten one more. It's just the same as saying four lots of three is the same as three lots of four. Or, as part of your multiplication tables, $4 \times 3 = 3 \times 4$.

## 4. Match Puzzles

*Squares from 24 Matches*
(*a*) One extra square in two ways

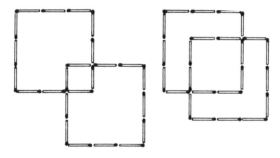

(*b*) Four extra squares—seven in all

(c) Seven squares

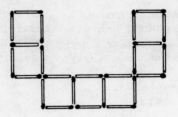

Eight squares with one extra, larger square

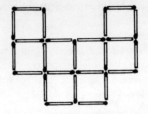

Eight squares with two extra, larger squares

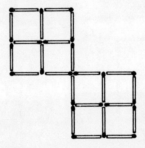

Nine squares with five extra squares

*Half-Match Squares*
Yes. Four larger squares.

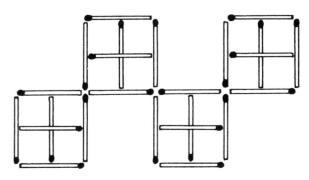

*Third-Match Squares*
Yes. Fifteen larger squares.

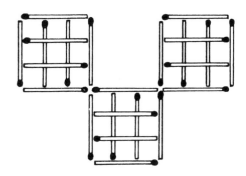

*Fifth-Match Squares*
Yes. 60 larger squares.

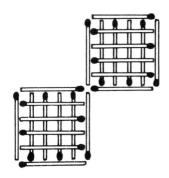

## Move-or-Remove Puzzles I

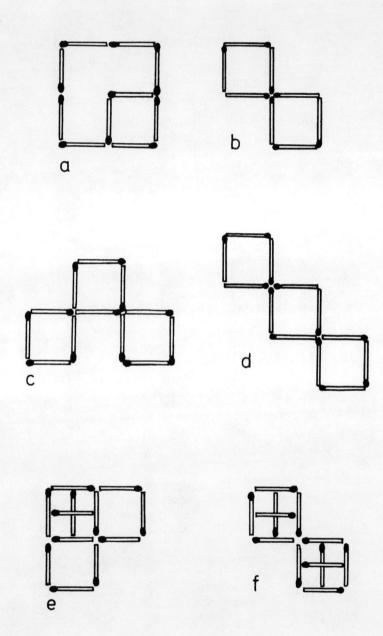

*Move-or-Remove Puzzles II*
(*a*) Remove the 12 matches inside the large square, and use them to make another large square.

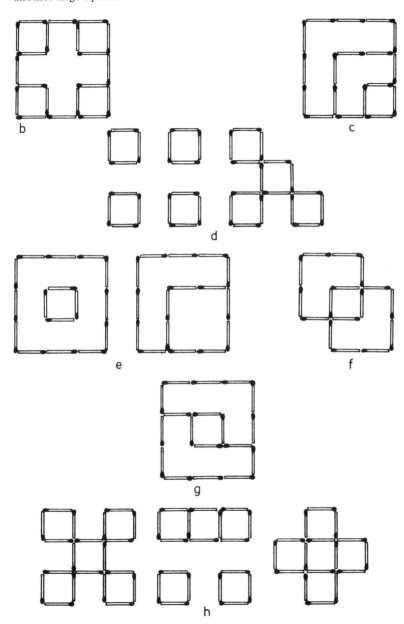

*Windows*

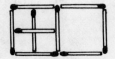

*Greek Temple*

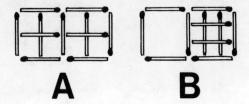

**A**   **B**

*An Arrow*

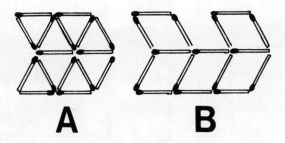

**A**   **B**

*Vanishing Trick*
There is one 4-by-4 square, four 3-by-3 squares, and nine 2-by-2 squares, and, of course, 16 small squares, making 1 + 4 + 9 + 16, or 30 squares in all.

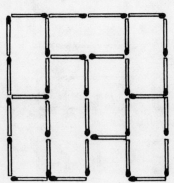

*Take Two*

*Six Triangles*

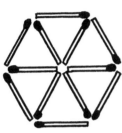
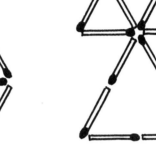

*Squares and Diamonds*

*Stars and Squares*

*A Grille*

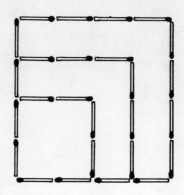

*The Five Corrals*

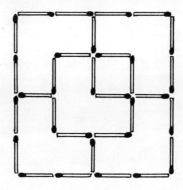

*Patio and Well*

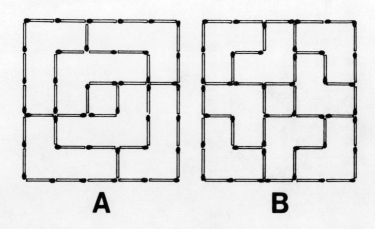

*Four Equal Plots*

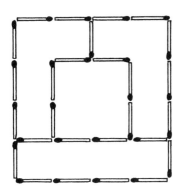

*Get Across the Pool*

*Spiral into Squares*

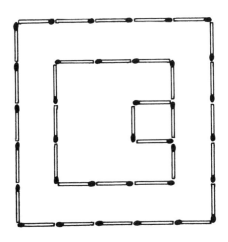

*More Triangle Trickery*
(*a*) Move 3 matches in the square corner as shown to form a step. Its area is 2 square match units less than that of the original triangle (6 square match units); thus, it is 4 square match units.

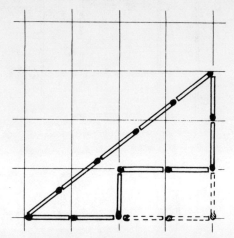

(*b*) Move 2 more matches to make another step. Its area is then 3 square match units less—that is, its area is 3 square match units.

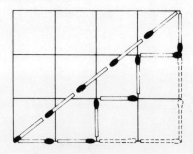

*Triangle Trio*

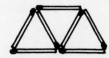

*Triangle Quartet*
The answer is a triangular pyramid, a *tetrahedron*.

*3 Times the Area*

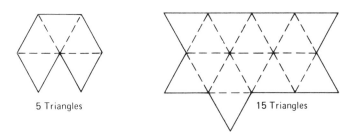

5 Triangles     15 Triangles

*Cherry in the Glass*
Slide one match across and move the other like this:

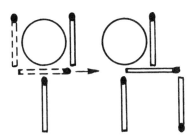

## 5. Coin and Shunting Problems

*Coin Sorting in Pairs*
We have numbered the coins to explain the answer. The coins can be regrouped in three moves: Move coins 1 and 2 two places to the left. Fill the gap by 4 and 5. Jump 5 and 3 over to the far left.

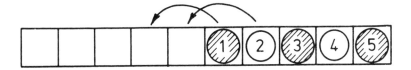

*Rats in a Tunnel*
The eight moves are as shown. There are two general rules: (1) Shift a coin forward into a free space, then jump another coin over the coin just shifted. (2) Always make shifts or jumps into the center of the tunnel first *before* making jumps or shifts away from it.

A wrong move 3 is shown to indicate how you can get blocked. The rats should end up exactly exchanged and not with two spaces between each of the black rats or each of the white ones.

| | | | | | | |
|---|---|---|---|---|---|---|
| Start | ● | ● | | | ○ | ○ |
| Move 1 | ● | | →● | | ○ | ○ |
| 2 | ● | ○ | ● | | | ○ |
| 3 | ● | ○ | | →● | | ○ | Blocked! (Move inwards before moving outwards.) |
| (Instead do this) 3 | ● | ○ | ● | ○← | | |
| 4 | ● | ○ | | ○ | ● | |
| 5 | | ○← | ● | ○ | ● | |
| 6 | ○← | | ● | ○ | ● | |
| 7 | ○ | ○← | ● | | ● | |
| 8 | ○ | ○ | →● | ● | | Done! |

*Three-Coin Trick*
(a) Using H for heads and T for tails, the moves are:

| Begin | T | H | T |
|---|---|---|---|
| Move 1 | H | T | T |
| Move 2 | H | H | H |

Done!
(b) No, it cannot be done. Each move is not going to alter whether there is an even or an odd number of tails (or heads). As you see above, at each stage there is always an odd number of heads and an even number of tails. So you cannot get three tails because 3 is an odd number.

*Triangle of Coins*
The trick is to move the coins in the opposite way to which you want the final triangle to point.

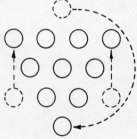

## Five-Coin Trick

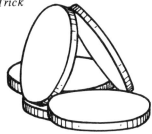

## Five-Coin Puzzle
The general plan is as follows. You can shorten the number of moves but this description is easy to remember.

Slide all five coins around clockwise till the half-dollar is in the top right corner (picture 1). You note there is now a space between the half-dollar and the penny. Here we break the flow of coins. This is the cunning bit. Shift around *just* the penny, the dimes and the nickel in a clockwise direction until the penny is in the bottom left corner (picture 2). Shift around *just* the dime, the half-dollar, and the quarter in a counterclockwise direction until the half-dollar is next to the penny (picture 3). All you have to do now is slide all five coins around clockwise until the penny is just above the half-dollar, on the left.

The trick was to split into two the flow of coins and reverse the direction of flow of three of them.

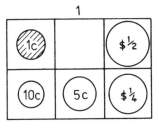

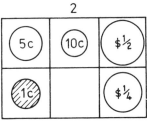

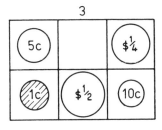

## Coin Changeovers
The pennies and nickels *can* change places in both cases.

## Mission Impossible?
Writing $F$ for Dr. Fünf and $S$ for Dr. Sieben, and 5 and 7 for the agents, here is one way of completing the mission. $F$, 5, $S$, and 7 start on the Slobodian bank. First, $F$ and 5 row across the river; $F$ stays on far bank. 5 rows back, picks up fellow agent 7 and rows him over, leaving $S$ alone on the Slobodian bank. Then 7 rows back alone and picks up $S$ and takes him across to join the other two. Mission *possible!*

*Railroad Switch*
Here are the six moves:

*(1)* The engine driver moves past *B*, backs up into *BC*, and couples on the black car.
*(2)* He pulls the black car past *B*, then he backs up into *AB* and uncouples the black car. Then he moves past *B* and backs into *BC* again.
*(3)* He backs past *C* and then shunts forward into *AC* and couples up the white car.
*(4)* He pushes the white car onto the main line out past *A*. Still coupled to it, he backs up along *AB* and couples on the black car; he is now sandwiched between the two cars.
*(5)* Sandwiched between the two cars, he backs down past *B*. Then he moves up *BC*, where he uncouples the white car.
*(6)* He now backs past *B* and then moves forward past *A*, still towing the black car. He then backs up *AC* and uncouples it. He moves out of *AC* past *A*; then he backs up into the stretch *AB*. He is now facing the other way.

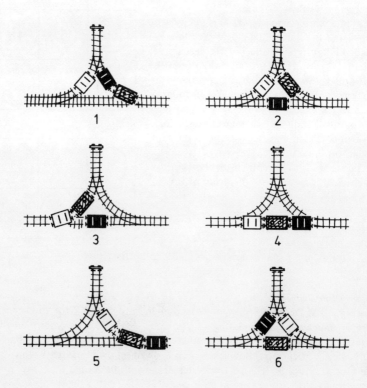

*Restacking Coins*
Just seven moves are needed, so they shouldn't have taken *too* long to do! Here is the relationship between the number of moves and the number of coins:

| Coins | 1 | 2 | 3 | 4 | 5 |
|---|---|---|---|---|---|
| Moves | 1 | 3 | 7 | 15 | 31 |

The number of moves is 2 times itself the same number of times as coins used, less 1. Thus for three coins, it is $(2 \times 2 \times 2) - 1$, or $8 - 1$ which is 7.

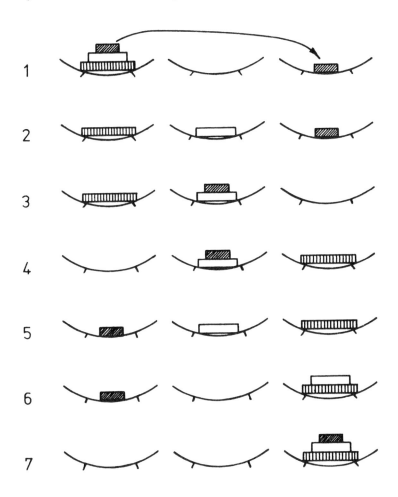

*River Crossing*

First, the two boys cross in the boat:

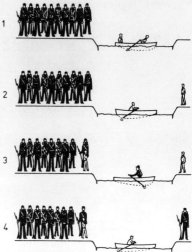

Now one soldier is across the river and the two boys and the boat are on the first bank with the soldiers. Repeat the operation as many times as there are soldiers! You note that the number of soldiers wasn't given; it doesn't matter.

*Collision Course?*

- (1) W (white engine) with its car backs far out to the right (one reversal).
- (2) W runs onto the switch, leaving its car on main track.
- (3) B (black engine) with its car runs out to the right.
- (4) W backs onto main track (two reversals).
- (5) W couples with black car and moves forward to left of switch.
- (6) B backs onto switch (three reversals).
- (7) W and black car back off to right and couple with white car (four reversals).
- (8) W pulls two cars to left of switch.
- (9) B runs onto main track.
- (10) B backs to train (five reversals).
- (11) B picks up two cars and pulls them to right.
- (12) B backs rear (black) car onto switch (six reversals).
- (13) B pulls one car to right.
- (14) W backs past switch and picks up white car from engine (seven reversals).
- (15) W pulls its car off to left and away.
- (16) B backs up to switch and picks up its own car (eight reversals).
- (17) B engine pulls its car from switch onto track and goes on its way.

# 6. Reasoning and Logical Puzzles

*Thinking Blocks*
**A.**

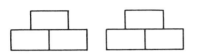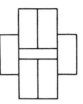

**B.**

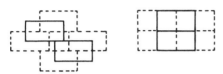

**C.**

**D.**

*Martian Orders!*
(*a*) Thalia, Zane, Xeron; (*b*) Sheree, Thalia, Zane, Xeron.

*What Shape Next?*
(*A*) Shape *c*, (*B*) shape *e*.

*IQ Puzzle*
Shape 3.

*Odd Shape Out*
*a*, 2; *b*, 3; *c*, 3; *d*, 1.

*The Same Shape*
(*A*) Shape *d*, (*B*) shape *c*.

*Next Shape, Please*

*The Apt House*
House 2.

*Who Is Telling the Truth?*
Con is.

*The Colored Chemicals Puzzle*
Orange, yellow, and green are poison.

*Mr. Black, Mr. Gray, and Mr. White*
The key is that the man in white is talking to Mr. Black and so cannot *be* him. Nor can he be Mr. White, since nobody is wearing his own color. So the man in white must be Mr. Gray. We can show what we know like this:

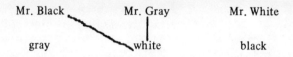

The straight line shows what must be true; the wiggly line shows what cannot be true. Mr White cannot be wearing white; so he's in black. That leaves Mr. Black wearing gray.

*Hairdresser or Shop Assistant?*
Amy and Babs are shop assistants; Carol is a hairdresser.

*The Zookeeper's Puzzle*
Art and Cora.

*Who's Guilty?*
Alf and Bert are guilty.

*Who's in the Play?*
Charles and Alice.

*Tea, Coffee, or Malted Milk?*
Malted milk.

*Soda or Milkshake?*
Suppose Alan has a soda. Then (*a*) says Bet has a milkshake. But (*c*) tells you Cis cannot then have a milkshake and must have a soda. But (*b*) says both Alan and Cis cannot both have a soda. Alan cannot have a soda; so Alan has a milkshake. From (*b*) that means Cis has a soda. Which, from (*c*), leaves Bet free to choose either a soda or a milkshake. Thus there are two possible orders: (1) Alan, milkshake; Cis, soda; and Bet, soda. (2) Alan, milkshake; Cis, soda; Bet, milkshake.

*Newton's Kittens*
Obviously the kittens could have gotten in and out by the same hole as the mother cat!

*March Hare's Party*
Sylvie had tea under the tree at table 1 because she wouldn't go near water. Al and Barbra sat at table 3: He couldn't take her in the boat to table 4. Gary joined them at 3, roller-skating over the bridge: He couldn't go to table 2 because of the "no boys" rule. That leaves Don, who wouldn't sit with Gary at table 3 and also couldn't take the path to table 2. Don rowed to table 4 and sat by himself.
   Answer: table 1, Sylvie; 2, nobody; 3, Al, Barbra, and Gary; and 4, Don.

*Marriage Mix-up*
Ted is married to Barbra with daughter Ruth, Pete to Sue with daughter Wendy, and Charlie to Nicola with son Dick. The reasoning goes like this: Ted's daughter is not Wendy. So his daughter must be Ruth. So Pete is father of other girl, Wendy. Which means Charlie is father of Dick. So his wife cannot be Barbra because she has a daughter. (Assume a girl plays Annie and Ophelia!) His wife is not Sue, so his wife has to be Nicola. Now Pete's daughter is not Barbra's daughter because they have only one child each. So Pete cannot be married to Barbra. That means Ted is married to Barbra, and Pete therefore to Sue.

*Who Does Which Job?*
Orville is bartender and singer; Virgil is private eye and racing driver; Homer is jockey and cardsharp. This is how you get the answer. Draw a table to show the men and the jobs, and fill it in as follows:

| Facts used | | Orville | Virgil | Homer |
|---|---|---|---|---|
| 3 | Private eye | | | |
| 1  2 | Racing driver | | | X |
| 2 | Singer | | X | X |
| 3  4 | Jockey | X | X | |
| 4 | Bartender | | | |
| 1 | Cardsharp | | | |

First look at the *jobs*. Fact *1* tells us the bartender is not the same man as the racing driver. Put a *1* beside them (as shown). Similarly the racing driver (*2*) is not the singer (*2*). And so on. Now look at the *men*. Fact *2* tells us Homer is neither the racing driver nor the singer; so put an *X* in the table under *Homer* opposite those two jobs, as shown. Fact 5 says Virgil is not the singer; put an *X* under *Virgil* opposite *Singer*. Fact 6 tells you Homer is the jockey as Virgil and Orville are not; opposite *Jockey* put *X*'s under *V* and *O* and a check under *H*.

Now for the reasoning. Orville must be the singer—since neither Virgil nor Homer is—so put a check under *O* opposite *Singer*. Here is the table so far in brief:

|   | O | V | H |
|---|---|---|---|
| P |   |   |   |
| R |   |   | X |
| S | ✓ | X | X |
| J | X | X | ✓ |
| B |   |   |   |
| C |   |   |   |

To fill the *Singer* line put a check under Orville (*O*). Then in the *Jockey* line put a check under Homer (*H*). And so on.

Fact 2 tells us to put a cross under *H* opposite R. Fact 3 gives an *X* under *H* opposite P and B. That leaves only C for *H*'s second job: put a check there. The table looks like this:

|   | O | V | H |
|---|---|---|---|
| P |   |   | X |
| R |   |   | X |
| S | ✓ | X | X |
| J | X | X | ✓ |
| B |   |   | X |
| C |   |   | ✓ |

Now we can put a check under *V* opposite R. So Fact *1* gives an *X* under *V* opposite B—thus forcing a ✓ under *O* on that line (that is, Orville's second job is bartender). Finally, the bottom line with two *X*'s means Virgil's second job is P (private eye).

*Birds and Insects*
Answer *A* alone follows logically.

*Wonderland Golf*
Five shots: DDDSD or SDDSD. There *is* a pattern. To see it, turn back to the map of the golf course. Working backwards from hole 18, a D shot gets you back to hole 9, then an S shot to 8, followed by three D's to the first tee. (From 2 to 1 could be an S as well.) You can also work it out by arithmetic. Divide the hole number by 2 over and over again, noting if there is a remainder of 1. For 18 you get:

$$\begin{array}{|c|} \hline 18 \\ \hline 9 \\ 4 \; r \; 1 \\ 2 \\ 1 \\ \hline \end{array}$$

Count up the number of answers and remainders: 9, 4, 1, 2, 1, which makes five numbers; that's how many shots it takes. This rule works for *any* hole.

*Mad Hatter's Tea Party*
Set $G$ to table 1, $M$ to 3, and $B$ to 4. Table 2 stays empty.

## 7. Mathematical Games

*Nim*
The way to calculate a winning position is best shown with the starting position of Nim. It has 3, 4, and 5 matches. We rewrite the rows in binary— that is, in powers of 2, or in "doublings." The numbers 100, 10, and 1 in binary are 4, 2, and 1 in everyday numbers. While *11* in usual counting numbers means 1 ten and 1 one, in binary it means 1 two and 1 one. We can make 3 out of 2 + 1 and in binary write it as 11. Then 4 in binary is 100, meaning 1 four and no twos and no ones, and 5 in binary is 101, meaning 1 four, no twos and 1 one. We set the rows out in binary as follows:

|            | Matches | Fours | Twos | Ones |
|------------|---------|-------|------|------|
| Top row    | 3       |       | 1    | 1    |
| Middle row | 4       | 1     | 0    | 0    |
| Bottom row | 5       | 1     | 0    | 1    |
| Totals     |         | 2     | 1    | 2    |

As you see, we added each column up in ordinary numbers; but we did not "carry" numbers over from one column to the next. Two column totals are even, and one, the middle column, is odd. To make the position safe for yourself, all you do is make the totals of each column *even*. So your first move is to take 2 matches from the top row, as explained. This changes the top binary number to 1. The columns then become:

|  | Matches | Fours | Twos | Ones |
|---|---|---|---|---|
| Top row | 1 |  |  | 1 |
| Middle row | 4 | 1 | 0 | 0 |
| Bottom row | 5 | 1 | 0 | 1 |
| Totals |  | 2 | 0 | 2 |

Now each column adds up to an even number. The position is safe.

*Daisy*
Here is the second player's winning strategy: Say the first player takes one petal; then the second player takes two petals, which must be next to each other, directly opposite the one taken by the first player. If the first player takes two adjacent petals, the second player takes one petal, again directly opposite. Either way this leaves two sets of five petals, symmetrically arranged about the two spaces. All the second player has to do now is to keep the pattern symmetrical, taking special note of the spaces.

*The Cop and the Robber*
The cop has first to go around the triangular block at the top left corner. Then he is an odd number of corners away from the robber and can catch him—provided the robber does not go around the triangle! Remember, there are only three corners in the triangular block, and you can get right around it in three moves.

*Morra*

Morra's Winnings

|  |  | Other shows | |
|---|---|---|---|
|  |  | 1 finger | 2 fingers |
| Morra shows | 1 finger | + 2 | − 3 |
|  | 2 fingers | − 1 | + 2 |

Morra's best strategy, to reduce his losses, is to show two fingers all the time; then he never loses more than one penny.